Data-Centric Biology

Data-Centric Biology: A Philosophical Study

Sabina Leonelli

The University of Chicago Press :: Chicago and London

The University of Chicago Press, Chicago 60637
The University of Chicago Press, Ltd., London
© 2016 by The University of Chicago
All rights reserved. Published 2016.
Printed in the United States of America

25 24 23 22 21 20 19 18 17 16   1 2 3 4 5

ISBN-13: 978-0-226-41633-5 (cloth)
ISBN-13: 978-0-226-41647-2 (paper)
ISBN-13: 978-0-226-41650-2 (e-book)
DOI: 10.7208/chicago/9780226416502.001.0001

Library of Congress Cataloging-in-Publication Data

Names: Leonelli, Sabina, author.
Title: Data-centric biology : a philosophical study / Sabina Leonelli.
Description: Chicago ; London : The University of Chicago Press, 2016. |
    Includes bibliographical references and index.
Identifiers: LCCN 2016015882 | ISBN 9780226416335 (cloth : alk. paper) | ISBN
    9780226416472 (pbk. : alk. paper) | ISBN 9780226416502 (e-book)
Subjects: LCSH: Biology—Data processing—Philosophy. | Biology—
    Research—Philosophy. | Knowledge, Theory of. | Biology—Research—
    Sociological aspects. | Research—Philosophy.
Classification: LCC QH331.L5283 2016 | DDC 570.285—dc23 LC record available
    at https://lccn.loc.gov/2016015882

♾ This paper meets the requirements of ANSI/NISO Z39.48-1992 (Permanence
of Paper).

Contents

Introduction

Over the last three decades, online databases, digital visualization, and automated data analysis have become key tools to cope with the increasing scale and diversity of scientifically relevant information. Within the biological and biomedical sciences, digital access to large datasets (so-called big data) is widely seen to have revolutionized research methods and ways of doing science, thus also challenging how living organisms are researched and conceptualized.[1] Some scientists and commentators have characterized this situation as a novel, "data-driven" paradigm for research, within which knowledge can be extracted from data without reliance on preconceived hypotheses, thus spelling the "end of theory."[2] This book provides a critical counterpoint to these ideas by proposing a philosophical framework through which the current emphasis on data within the life sciences, and its implications for science as a whole, can be studied and understood. I argue that the real source of innovation in current biology is the attention paid to data handling and dissemination practices and the ways in which such practices mirror economic and political modes of interaction and decision making, rather than the emergence of big data and associated methods per se. We are not witnessing the birth of a data-driven method but rather the rise of a data-centric approach to science, within which efforts to mobilize, integrate, and visualize data are valued as contributions to discovery in their own right and not as a mere by-product of efforts to create and test scientific theories.

The main thesis of this book is that the convergence of digital technologies for the production, dissemination, and analysis of data and novel regulatory and institutional regimes is provoking a reshuffling of priorities in research practices and outcomes, with important consequences for what may be viewed as scientific knowledge and how that knowledge is obtained, legitimated, and used. The rise of data centrism has brought new salience to the epistemological challenges involved in processes of data gathering, classification, and interpretation and the multiplicity of conceptual, material, and social structures in which such processes are embedded. To document this phenomenon, I examine the processes involved in aggregating, mobilizing, and valuing research data within the life sciences, particularly the ways in which online databases are developed and used to disseminate data across diverse research sites. I use these empirical insights to develop a relational account of the nature and role of data in biological research and discuss its implications for the analysis of knowledge production processes in science more generally. The book thus addresses three issues of central concern to the philosophy, history, and social studies of science, as well as contemporary science and science policy: what counts as data in the digital age, and how this relates to existing conceptions of the role and use of evidence in the life sciences and elsewhere; what counts as scientific knowledge at a time of significant technological and institutional change, and how this relates to the social worlds within which data are produced, circulated, and used; and under which conditions large datasets can and should be organized and interpreted in order to generate knowledge of living systems.

There is nothing new or controversial to the idea that data play an important role in research. Indeed, the sophisticated methods developed to produce and interpret data are often viewed as demarcating science from other types of knowledge. Until recently, however, scientific institutions, the publishing industry, and the media have portrayed data handling strategies as conceptually uninteresting. While many scientists stress their importance in interviews or personal accounts, this is not the sort of activity that typically results in Nobel prizes, high-level publications, or large amounts of research funding. The best recognized contributions to science, and thus the currency for academic promotions and financial support, consist of written texts that make new claims about the world. Most philosophers of science have accepted and even fostered this approach by positing theories and explanations as the key outcomes of science, whose validation depends on inference from data, and paying little attention to processes of data production, mobilization, and analysis. Partly as a result of this perception, data handling practices have largely been delegated to laboratory technicians and archivists—"support staff" who are often not acknowledged or rewarded as direct contributors to the creation of knowledge. The existence of data that do not support the claims being made also tends to be disregarded in this system, since peer review typically checks whether the data submitted by authors constitutes satisfactory evidence for their claims but not whether other data generated by the same group could be viewed as counterevidence or as evidence for different claims.

This theory-centric way of thinking is being challenged by the emergence of computational technologies for the production, dissemination, and analysis of data and related regimes of funding and research assessment. This is particularly evident in molecular biology, where high-throughput technologies such as next-generation genome sequencing, microarray experiments, and systems to track organisms in the field have enormously increased scientists' ability to generate data. Hence many researchers produce vast quantities of data in the hope that they might yield unexpected insights, a situation that Ulrich Krohs has aptly dubbed "convenience experimentation."[3] In addition, as documented by Geoff Bowker, among others, computing and information technology have improved scientists' "memory practices" by enhancing their ability to store, disseminate, and retrieve data.[4] At last, in principle, data acquired within one project can now be shared through the Internet with other research groups, so that not only the original data producers but also others in the scientific community might be able to analyze those data and use them as evidence for new claims. Funding bodies, scientific institutions, national governments, and researchers are thus increasingly viewing data as scientific achievements that should be valued and rewarded independently of their immediate worth as evidence for a given hypothesis—because they could contribute to future research in multiple and unpredictable ways, depending on the type of analysis to which they are subjected.

This sentiment is eloquently captured by the editor of the journal *F1000 Research*, which was founded with the explicit aim to foster data dissemination: "If you have useful data quietly declining in the bottom of a drawer somewhere, I urge you to do the right thing and send it in for publication—who knows what interesting discoveries you might find yourself sharing credit for!"[5] This quote epitomizes the prominent status acquired by data in scientific practice and discourse and the normative strength acquired by the requirement to make research data freely available to peers. This is most visibly showcased by the emergence of the Open Science movement, which advocates the free and widespread circulation of all research components, including data; the eagerness with which funding bodies and governments are embracing data-sharing policies and developing related systems of implementation and enforcement; and the ongoing transformation of the publishing industry to improve quality and visibility for data publication— including the creation of journals devoted solely to documenting data dissemination strategies (e.g., *GigaScience*, started in 2012, and *Scientific Data*, launched in 2013).[6] As Stephen Hilgartner noted already in 1995, these developments signal the rise of a data-centric communication regime in science and beyond. In this book, I document these developments over the last three decades within the life sciences and situate them within the longer historical trajectory of this field, thus highlighting both the continuities and the ruptures that data centrism brings with respect to other normative visions for how research should be carried out and with which outcomes.

In particular, I focus on the experimental research carried out on model organisms such as fruit flies, mice, and thale cress within the second half of the twentieth century and the ways in which these efforts have intersected with other areas of biological research. Model organisms are nonhuman species used to research a wide range of biological phenomena in the hope that the resulting knowledge will be applicable to other species. Work on these organisms encompasses the vast majority of experimental efforts in biology, focusing particularly on molecular biology but also including studies of cells, tissues, development, immune system, evolutionary processes, and environmental interactions, with a view to enhance interdisciplinary understandings of organisms as complex wholes. It is one of the best-funded research areas in contemporary academia, and its development is intertwined with broader shifts in scientific research, public cultures, national policies, and global financial and governance systems—a multiplicity of accountabilities and interests that significantly affects the status and handling of data. Furthermore, it is highly fragmented, encompassing a wide variety of epistemic cultures, practices, and interests and multiple intersections with other fields, ranging from computer science to medicine, statistics, physics, and chemistry. The norms, instruments, and methods used to produce and evaluate data can therefore vary enormously, as do the types and formats of objects that biologists working on model organisms regard as data, which encompass photographs, measurements, specimens of organisms or parts thereof, field observations, experiments, and statistical surveys. This pluralism creates serious obstacles to any attempt to disseminate data beyond their original site of production. Most interestingly for my purposes, biologists have long recognized these obstacles and have an illustrious history of ingenuous attempts to overcome them. Multiple technologies, institutions, and procedures have emerged to facilitate the collection, preservation, dissemination, analysis, and integration of large biological datasets, including lists, archives, taxonomies, museum exhibits and collections, statistical and mathematical models, newsletters, and databases.[7] As a result, biologists have developed sophisticated labeling systems, storage facilities, and analytic tools to handle diverse sources of data, which makes the life sciences into an excellent case for exploring the opportunities and challenges involved in circulating data to produce knowledge.

This book explores these opportunities and challenges both empirically and conceptually. The first part of the book proposes an empirical study of what I call *data journeys*: the material, social, and institutional circumstances by which data are packaged and transported across research situations, so as to function as evidence for a variety of knowledge claims. Chapter 1 explores the technical conditions for data travel within model organism biology, focusing on the rise of databases to disseminate the data accumulated on these species. It reviews the structure of these databases, the labor involved in their development, and the multiplicity of functions that they play for their users and sponsors, and it also reflects on the significance of my use of travel metaphors to analyze these features. Chapter 2 locates these scientific efforts in a broader social and cultural context, documenting the emergence of institutions and social movements aiming to promote and regulate data travel so as to ensure that data maximize their value as evidence for knowledge claims. Attributions of scientific value are shown to be closely intertwined with attributions of political, economic, and affective value to data, which in turn illustrates the close interplay between research and the political economy of data centrism. I also demonstrate that what counts as data varies considerably depending on how data are handled and used throughout their journeys, and how this variation is managed affects the development and results of scientific inquiry.

Building on these insights, the second part of the book analyzes the *characteristics of data-centric biology*, particularly the role of data, experimental know-how, and theories in this approach to research. Chapter 3 reviews existing philosophical treatments of the status and uses of data in science and proposes an alternative framework within which data are defined not by their provenance or physical characteristics but by the evidential value ascribed to them within specific research situations. This is a relational view that makes sense of the observation that the circumstances of travel affect what is taken to constitute data in the first place. Within this view, the function assigned to data determines their scientific status, with significant epistemological and ontological implications. Chapter 4 expands on this view by considering the various forms of nonpropositional, embodied knowledge ("know-how") involved in making data useable as evidence for claims and the difficulties encountered when attempting to capture such knowledge in databases. This brings me to reflect on the nature of the reasoning involved in data-centric knowledge production, which is distributed among the individuals involved in different stages of data journeys. Chapter 5 focuses instead on the propositional knowledge underlying the travel of data, particularly the classification systems and related practices used to make data searchable and retrievable by database users. These practices demonstrate that data journeys are not simply theory-laden but can actually generate theories that guide data analysis and interpretation.

The third and final part of the book reflects on the *implications of data centrism*. Chapter 6 considers implications for biology, where data handling strategies can affect which data are disseminated and integrated; how, where, and with which results; as well as the visibility and future development of specific research traditions. I also reflect on what this means for an overarching understanding of data-centric science, particularly its historical novelty as a research mode. Chapter 7 examines implications for philosophical analyses of processes of inquiry, particularly for conceptualizations of the conditions under which research takes place. I critique the widespread use of the term "context" to separate research practices from the broader environment in which they take place and instead propose to adopt John Dewey's notion of "situation," which better highlights the dynamic entanglement of conceptual, material, social, and institutional factors involved in developing knowledge and clearly positions research efforts in relation to the publics for whom such knowledge is expected to be of value.

This brief outline shows how the book moves from the concrete to the abstract, starting from a conceptually framed analysis of particular data practices and culminating in a general perspective on data centrism and the role of data in scientific research as a whole, which I summarize in my conclusion. Focusing on specific realms of scientific activity does not hamper the depth or the breadth of philosophical analysis but rather grounds it on a concrete understanding of the challenges, concerns, and constraints involved in handling and analyzing data.[8] This way of proceeding embodies a scholarly approach that I like to call *empirical philosophy of science*, whose goal is to bring philosophical concerns and scholarship to bear on the daily practice of scientific research and everything that such practice entails, including processes of inquiry, material constraints, institutional settings, and social dynamics among participants.[9] To this aim, my analysis builds heavily on historical and social studies of science, as well as interactions and collaborations with relevant practitioners in biology, bioinformatics, and computer science. The methods used in this work range from argumentation grounded in relevant philosophical, historical, anthropological, and sociological literature to analyses of publications in natural science journals; con-

sultation of archives documenting the functioning and development of biological databases; and multisited ethnographic explorations, on- and offline,
of the lives and worlds that these databases create and inhabit. Between 2004
and 2015, I participated in meetings of curatorial teams and steering committees overseeing databases in model organisms biology. I witnessed—and
sometimes contributed to[10]—scientific and policy debates about whether
and how these and other research databases should be maintained, updated,
and supported in the future. I also attended and organized numerous conferences, training events, and policy meetings around the globe (including the
United Kingdom, France, Belgium, Italy, the Netherlands, Germany, Spain,
the United States, Canada, South Africa, China, and India) in which database users, publishers, funders, and science studies scholars debated the usefulness and reliability of these tools.

These experiences were essential to the writing of this book in three
respects. First, they helped me acquire a concrete sense of the challenges
confronted by the researchers involved in setting up and maintaining biological databases, which in turn led me to move away from the promissory—
and often unrealistic—discourse associated to "big data" and "data-driven
methods" and focus instead on the practical issues and unresolved questions raised by actual attempts to make data travel. Second, they served as
a constant reminder of the impossibility to divorce the analysis of the epistemic significance of data handling practices from the study of the conditions under which these practices occur, including the characteristics of the
scientific traditions involved; the nature of the entities and processes being
studied; and the institutional, financial, and political landscape in which
research takes place. Third, they enabled me to present and discuss my ideas
with hundreds of researchers of varying seniority and experience, as well
as with publishers, editors, policy makers, activists, science funders, and
civil servants engaged in debates over the implementation of Open Science
guidelines, the sustainability of databases, and the significance of the digital
age for scientific governance.[11] In addition to providing helpful feedback,
these interactions made me accountable to both scientists and regulators,
which again helped to keep my analysis responsive to questions and problems plaguing contemporary academic research. This is a situation that I
regard as highly generative and desirable for a philosopher. In John Dewey's
words, "Philosophy recovers itself when it ceases to be a device for dealing
with the problems of philosophers and it becomes a method, cultivated by
philosophers, for dealing with the problems of men."[12]

I am aware that this approach is at odds with some parts of the philosophy of science, particularly with philosophical discussions carried out in
the absence of any information about—or even interest in—the material

and social conditions under which knowledge is developed. My research is not motivated by the desire to understand the structure and contents of scientific knowledge in the abstract. Rather, I am fascinated by the ways in which scientists transform the severe and shifting constraints posed by their institutional location, social networks, material resources, and perceptual capabilities into fruitful opportunities to understand and conceptualize the world and themselves as part of it. My account builds on the work of philosophers who share my interest in the ingenuity, serendipity, and situatedness of research and thus disregards purely analytic discussions of topics such as confirmation, evidence, inference, representation, modeling, realism, and the structure of scientific theories. This is not because I regard such scholarship as irrelevant. Rather, it is because it is grounded in the presupposition that data analysis follows logical rules that can be analyzed independently of the specific circumstances in which scientists process data. My research experiences and observations are at odds with such an assumption and thus proceed in a different direction. While I hope that future work will examine the relation between the view presented here and the vast analytic scholarship available on these issues, my concern here is to articulate a specific philosophical account and its empirical motivations. Relating this account to other types of philosophical scholarship, both within the analytic and continental traditions, would require a completely different book and it is not my ambition to fulfill this mandate in this text.

One question I was asked over and over again while presenting this work to academic audiences was whether this research should be seen as a contribution to philosophy, science studies, or science itself. Is my assessment of data centrism intended to document the concerns of the biologists involved, thus providing an empirically grounded description of the state of the field? Is it rather a critical approach, aiming to place biological practices in a broader political, social, and historical context? Or is it a normative account of what I view as the conceptual, material and social foundations of this phenomenon, as typically offered by philosophers? I think of my account as attempting to encompass all three of these dimensions. It is intended to be, first and foremost, a philosophical assessment of data centrism and thus a normative position that reflects my perspective on what this phenomenon consists of, its significance for scientific epistemology, and how it relates to the broader spectrum of activities and methods associated with scientific knowledge production. At the same time, my long-term engagement with the scientific projects that I am discussing, as well as research on the historical and social circumstances in which these practices emerged and are currently manifested, have influenced and often challenged my philosophical views, leading to a position that is normative in kind and yet relies on what anthropologists call a "thick description" of my field of inquiry (i.e., a description that contains a specific, hard-won and situated interpretation of a given set of observations, rather than pretending to capture facts in a neutral, objective fashion).[13]

Many scientists and scientific institutions have referred to the beginning of the twenty-first century as a time of epochal change in how science is done. This book attempts to articulate what that change consists of, how deeply it is rooted in twentieth-century scientific practice, and what implications it has for how we understand scientific epistemology. Its attention to the specificity of data practices in biology is also its limit. I am proposing a framework that can serve as a starting point for studying how data are handled in other areas, as well as how biology itself will develop in the future—particularly since data centrism is increasingly affecting important subfields such as evolutionary, behavioral, and environmental biology, which I did not analyze in detail here. As I have tried to highlight throughout the text, my analysis is colored by my intuitions and preferences, as well as my perception of the specific cases that I have been investigating. This is the beauty and the strength of an explicitly empirical approach to the philosophy of science: like science itself, it is unavoidably fallible and situated and thrives on the joint efforts and disagreements of a diverse community of researchers.

Part One: Data Journeys

1   Making Data Travel: Technology and Expertise

On the morning of September 17, 2013, I made my way to the University of Warwick to attend a workshop called "Data Mining with iPlant."[1] The purpose of the workshop was to teach UK plant biologists how to use the iPlant Collaborative, a digital platform funded by the National Science Foundation in the United States to provide digital tools for the storage, analysis, and interpretation of plant science data. iPlant is a good example of the kind of technology, and related research practices, whose epistemic significance this book aims to explore. It is a digital infrastructure developed in order to make various types of biological data travel far and wide, so that those data can be analyzed by several groups of scientists across the globe, integrated with yet more data, and ultimately help biologists to generate new knowledge. While recognizing that gathering all existing plant data under one roof is a hopelessly ambitious goal, iPlant aims to incorporate as many data types—ranging from genetic to morphological and ecological—about as many plant species as possible. It also aims to develop software that plant biologists can easily learn to use for their own research purposes, thus minimizing the amount of specialized training needed to access the resource and facilitating its interactions with other digital services and databases.

The iPlant staff, which comprises over fifty individuals with expertise in both computer science and experimental biology, took a few years to get started on this daunting project. This is because setting up a digital infrastructure to support scientific inquiry involves tackling substantial challenges involving the collection, handling, and dissemination of data across a wide variety of fields, as well as devising appropriate software to handle user demands. Initially, iPlant staff had to determine which features of data analysis are most valued and urgently needed by plant scientists, so as to establish which goals to tackle and in which order. At the outset of the project in 2008, a substantial portion of funding was therefore devoted to consultations with members of the plant science community worldwide in order to ascertain their requirements and preferences. iPlant staff then focused on making these ideas practically and computationally feasible given the technology, manpower, and data collections at hand. They organized the physical spaces and equipment needed to store and manage very large files, including adequate computing facilities, servers powerful enough to support the operations at hand, and work stations for the dozens of staff and technicians involved across several campuses in Texas, California, Arizona, and New York. They also developed software for the management and analysis of data, which would support teams based at different locations and the integration of data of various formats and provenance. These efforts led to even more consultations with biologists (to check whether the solutions singled out by iPlant would be acceptable) as well as many groups involved in building the resource, such as the software developers, storage service providers, mathematicians, and programmers. The first version of the iPlant user interface, called Discovery Environment, was not released until 2011.

Plant scientists around the world watched these developments with anticipation in the hope of learning new ways to search existing data sets and make sense of their own data. The workshop at Warwick was thus well attended, as most plant science groups in the United Kingdom sent representatives to the meeting. It was held in a brand-new computer room situated at the center of the life sciences building—a typical instance of the increasing prominence of biological research performed through computer analysis over "wet" experiments on organic materials.[2] I took my place at one of the 120 large iMacs populating the room and set out to perform introductory exercises devised by iPlant staff to get biologists acquainted with their tools. After the first hour, I started to perceive some restless shuffling near me. Some of the biologists were getting impatient with the amount of coding and programming involved in the exercises and protesting to their neighbors that the data analysis they were hoping to carry out did not seem to be feasible within that system. Indeed, far from being able to use iPlant to their research advantage, they were getting stuck with tasks such as uploading their own data into the iPlant system, understanding which data formats worked with the visualization tools available in the Discovery Environment, and customizing parameters to fit existing research goals—and becoming frustrated as a result.

This impatience may appear surprising. These biologists were attending this workshop precisely to acquaint themselves with the programs used to power the computational tools offered by iPlant in anticipation of eventually contributing to their development—indeed, the labs had selected their most computationally oriented staff members as delegates for this event. Furthermore, as iPlant coordinators kept repeating throughout the day, those tools were the result of ongoing efforts to make the interface used for data analysis flexible to new uses and accessible to researchers with limited computer skills, and iPlant staff was at hand to help with specific queries and problems (in the words of iPlant co–principal investigator Dan Stanzione, "We are here to enable users to do their thing"). Yet I could understand the unease felt by biologists struggling with the limits and challenges of iPlant tools and the learning curve required to use them to their full potential. Like them, I had read the "manifesto" paper in which iPlant developers explained their activities, and I had been struck by the simplicity and power of their vision.[3] The paper starts with an imaginary user scenario: Tara, a biologist interested in environmental susceptibility of plant genomes, uses iPlant software to seamlessly integrate data across various formats, thousands of genomes, and hundreds of species, which ultimately enables her to identify new patterns and causal relations between key biological components and processes. This example vividly illustrates how data infrastructure could help understand how processes at the molecular level affect, and are in turn affected by, the behavior, morphology, and environment of organisms. Advances in this area have the potential to foster scientific solutions to what Western governments call the "grand challenges" of our time, such as the need to feed the rapidly increasing world population by growing plants more efficiently. As is often the case with large data infrastructures set up in the early 2000s, the stakes involved in the expectations set up by iPlant are as high as they can be. It is understandable that after reading that manifesto paper, biologists attending the workshop got frustrated when confronted with the challenges involved in getting iPlant to work and the limitations in the types of analyses that iPlant could handle.

This tension between promise and reality, between what data technologies can achieve in principle and what it takes to make them work in practice, is inescapable when analyzing any instance of data-centric research and

constitutes the starting point for my study. On the one hand, iPlant exemplifies what many biologists see as a brave new research world, in which the
billions of data churned out by high-throughput machines can be integrated
with experimentally generated data, leading to an overall understanding of
how organisms function and relate to each other. On the other hand, developing digital databases that can support this vision requires the coordination of diverse skills, interests, and backgrounds to match the wide variety
of data types, research scenarios, and expertises involved. Such coordination
is achieved through what I will call *packaging procedures* for data, which
include data selection, formatting, standardization, and classification, as
well as the development of methods for retrieval, analysis, visualization, and
quality control. These procedures constitute the backbone of data-centric
research. Inadequate packaging makes it impossible to integrate and mine
data, thus calling into question the plausibility of the promises made by
the developers of large data infrastructures. As exemplified by the lengthy
negotiations surrounding the development of iPlant, efforts to develop adequate packaging involve critical reflection over the conditions under which
data dissemination, integration, and interpretation can or should take place
and who should be involved in making them possible.

In this chapter, I examine the unresolved tensions, practical challenges,
and creative solutions involved in packaging data for dissemination.[4] I focus
on the procedures involved in labeling data for retrieval within model organism databases.[5] These databases constitute an exceptionally sophisticated
attempt to support data integration and reuse, which is rooted in the history
of twentieth-century life science, particularly the rise of molecular biology
in the 1960s and large-scale sequencing projects in the 1990s. In contrast
to infrastructures such as GenBank that only cater for one data type, they
are meant to store a variety of automatically produced and experimentally
obtained data and make them accessible to research groups with markedly
different epistemic cultures.[6] I explore the wealth and diversity of resources
that these databases draw on to fulfill their complex mandate and identify
two processes without which data could not travel outside of their original
context of production: decontextualization and recontextualization. I then
discuss how the introduction of computational tools to disseminate large
data sets is reconfiguring the skills and expertise associated with biological
research through the emergence of a new professional figure: the database
curator. Finally, I introduce the notion of data journeys and reflect on the
significance of using metaphors relating to movement and travel when examining data dissemination practices.

1.1 The Rise of Online Databases in Biology

In the period following the Second World War, biology entered a molecular bandwagon. Starting from the 1960s, biochemistry and genetics absorbed the vast majority of investments and public attention allocated to biology, culminating in the genome sequencing projects of the 1990s. These projects, which included the Human Genome Project and initiatives centered on nonhuman organisms such as the bacterium *Escherichia coli* and the plant *Arabidopsis thaliana*, were ostensibly aimed at "deciphering the code of life" by finding ways to map and document the string of nucleotides contained in the genome.[7] Many biologists cast doubt over their overall usefulness and scientific significance, and philosophers condemned the commitment to genetic reductionism that these projects seemed to support (namely, the idea that life can be understood primarily by reference to molecular processes by determining the order of nucleotides within a DNA molecule).[8] Despite these objections, these projects were very well funded and the rhetoric of coding and mapping life captured the imagination of national governments and media outlets, resulting in a public visibility rarely enjoyed by scientific initiatives.[9]

When many sequencing projects announced completion in the early 2000s, it became clear that they had indeed yielded little biological insight into the functioning of organisms, particularly when compared to the hype and expectations surrounding their initial funding. This fact in itself can be viewed as a significant outcome. It demonstrated that sequence data are an important achievement but do not suffice to yield an improved understanding of life. Rather, they need to be integrated with data documenting other biological components, processes, and levels of organization, such as, for instance, data acquired within cell and developmental biology, physiology, and ecology. Sequencing projects can thus be seen as spelling the end of genetic reductionism in biology and opening the door to holistic, integrative approaches to organisms as "complex wholes" such as those touted within systems biology.[10]

Another major outcome of sequencing projects was their success in bringing the scientific importance of activities of data production, dissemination, and integration to the attention of biologists, funding agencies, and governments.[11] Despite their central role throughout the history of biological research, practices of data collection and dissemination have not enjoyed a high status and visibility outside narrow circles of experts. Developing a smart archive or a way to store samples would typically be regarded as a technical contribution, rather than as a contribution to scientific knowledge, and only when those tools were used to generate new claims about the world would the word "science" be invoked. As I stressed in the introduction, this situation has changed over the last decades, and it is no coincidence that such a shift in the prominence of data practices has coincided with the elevation of sequencing from mere technology to a scientific specialty in its own right, requiring targeted expertise and skills.[12]

As highlighted by Hallam Stevens in his book *Life Out of Sequence*, the data produced by sequencing projects have come to exemplify the idea of "big data" in biology. It is in the context of sequencing projects that high-throughput machines, able to generate large quantities of data points with minimal human intervention, were developed. These projects produced vast data sets documenting the genotypes of organisms, which in turn fueled debate on how these data could be stored, whether they could be efficiently shared, and how they could be integrated with data that were not generated in a similarly automated manner. They generated interest in producing data about other aspects of subcellular biology, most prominently through "omics" data including metabolomics (documenting metabolite behavior), transcriptomics (gene expression), and proteomics (protein structure and functions). Furthermore, sequencing projects established a template for how the international research community could cooperate, particularly in the form of large-scale projects and networks through which access, use, and maintenance of machines and data infrastructures was structured and regimented.[13] Finally, sequence data became a classic example of biological data produced in the absence of a specific research question, as ways of exploring the molecular features of a given organism rather than testing a given hypothesis. More than their sheer size and speed of production, this disconnection between research questions and data generation is what marks sequences out as big data.[14] As in comparable cases in particle physics, meteorology, and astronomy, the acquisition of these data required large efforts and diverted investments from other areas of inquiry, and yet there was no certainty as to whether and how sequence data could be used to deliver the promised biological and medical breakthroughs—a situation that generated high levels of anxiety among funders and members of the scientific public, as well as a sense of urgency about finding ways to analyze and interpret the data.

This space of opportunity mixed with anxiety proved decisive to the development and scientific success of model organism databases. To explain how this came about, I need to briefly introduce the key characteristics of model organisms and reflect on their history as laboratory materials. Model organisms are a small number of species, including the fruit fly (*Drosophila melanogaster*), the nematode (*Caenorhabditis elegans*), the zebrafish (*Danio rerio*), the budding yeast (*Saccharomyces cerevisiae*), the weed (*Arabidopsis thaliana*), and the house mouse (*Mus musculus*), whose study has absorbed the vast majority of experimental efforts within biology (and particularly molecular biology) over the last sixty years.[15] Some of the reasons for this extraordinary success are practical. They are relatively small in size, highly tractable, and have low maintenance costs. They also possess biological traits—such as the transparent skin of zebrafish, which enable the observation of developmental processes without constant invasive interventions—that make them particularly useful for experimental research. In the words of Adele Clarke and Joan Fujimura, they are "the right tools for the job."[16] Most important for my present purposes, however, are the scientific expectations linked to focusing research on model organisms. It is typically assumed that insights obtained on their functioning and structure will foster the understanding of other species, ranging from relatively similar organisms (*Arabidopsis* generating insights into crop species, for instance) all the way to humans (as most obviously in the case of mice, which are routinely used as models for human diseases). This does not necessarily mean that model organisms are intrinsically better representations of biological phenomena than other species. Rather, the main reason for the success of these organisms lies in the way in which research on them has been managed and directed from the outset, particularly the interdisciplinary ambitions and collaborative ethos underlying their adoption as laboratory materials.

The biologists who pioneered the systematic use of model organisms in biology, including T. H. Morgan in the 1920s (fruit fly), Sydney Brenner in the 1960s (nematode), Maarten Koornneef and Chris Somerville in the 1970s (thale cress), and George Streisinger in the 1980s (zebrafish), shared a similar vision of how research should be conducted. They relentlessly promoted the sharing of ideas, data, and samples as a norm for scientific interactions, including at the prepublication stage—a remarkable feat, particularly in the context of the notoriously competitive culture characterizing biomedical research.[17] This was not simply the result of individual preferences and strong charisma: the wide and free dissemination of data was essential to the scientific success of their research programs, whose ultimate goal was to research organisms as complex wholes through an interdisciplinary approach that would include genetics as well as cell biology, physiology, immunology, morphology, and ecology. Proponents of model organism research believed that the best strategy to achieve such integrative understanding was to focus on a small number of species, explore as many aspects of their biology as possible, integrate the resulting data in order to obtain an overall understanding of their biology, and then use these results and the related infrastructure as reference points for the study of other species. This approach to research, widely supported by North American and European science funders and administrators in the 1980s and 1990s, was expected to provide a blueprint for how several branches of biology could be combined to understand organisms as complex wholes, which could then be used as a reference for the development of comparative, cross-species research. In joint work with Rachel Ankeny, I characterized this vision for how biological research should be conducted as a specific research strategy, in which data integration plays a central role and the expectation that data will be widely disseminated drives financial investment in data production and infrastructures for data sharing.[18] Within this view, model organisms can be characterized not in terms of their representative power with respect to other species or phenomena, as is often emphasized in historical and philosophical literature, but in terms of the research infrastructures, data collections, and biological knowledge available on them. In short, model organisms can be defined as organisms about which much is known; the knowledge of these creatures can be easily accessed and used in order to study other organisms.

The research strategy at the heart of model organism research has been widely critiqued, partly because of its emphasis on sequencing data and genetics (which made it vulnerable to antireductionist arguments) and partly because it sidesteps biodiversity and variation among organisms in ways that are arguably detrimental to biological understanding.[19] More often than not, biologists working on model organisms are aware of their limited representational power and do not consider them to be ideal exemplars of the biological processes that they wish to study. Rather, they are interested in the opportunities that these organisms offer as platforms to integrate results across disciplines and locations, an approach that has proved to be highly successful. The use of model organisms as "boundary objects" for multiple research traditions has generated key biological insights, ranging from an understanding of circadian rhythms (the daily and seasonal cycles through which organisms regulate their biological activities) to the ability to control the development of specific traits through genetic intervention.[20] It has brought genetic research in dialogue with developmental and evolutionary biology, resulting in some cases in the creation of new interdisciplinary communities. And, most importantly for my purposes, it has fostered the creation of some of the most sophisticated online databases in biology to date.

Indeed, given their commitment to data sharing and interdisciplinary integration, it should not come as a surprise that biologists working on model organisms played a key role in promoting sequencing projects and related data infrastructures in the 1990s. The accumulation of these data provided a common goal and unifying factor for model organism communities, as well as a way to channel, coordinate, and finance their efforts to store, circulate, and retrieve large masses of data. Biologists working on model organisms

played a key role in discussions over how to disseminate sequence data, leading to a consensus that these data should be made freely available without
restrictions, as formalized by the Bermuda Rules.[21] These biologists had a
relatively clear idea of what the usefulness of sequence data may be and how
it may be exploited to improve existing understandings of organisms. They
could therefore productively address the anxieties felt by funders and peers
concerning the future exploitation of sequence data. Furthermore, many of
them had already attempted to use digital repositories for the storage and
dissemination of genetic information and were eager to expand those efforts
to tackle larger amounts and varieties of data.[22] This conjunction of factors
resulted in major funding bodies, such as the National Science Foundation
in the United States and research councils in the United Kingdom, allocating
substantial funding to digital infrastructures for the dissemination of model
organism data in the late 1990s. Thus model organism databases emerged
with the immediate goal of storing and disseminating genomic data and
the longer-term vision of (1) incorporating and integrating any data available on the biology of the organism in question within a single resource,
including data on physiology, metabolism, and even morphology; (2) allowing and promoting cooperation with other community databases so
that the available data sets would eventually be comparable across species;
and (3) gathering information about laboratories working on each organism and the associated experimental protocols, materials, and instruments,
providing a platform for community building. Particularly useful and rich
databases included the ones dedicated to *Drosophila* (FlyBase), *C. elegans*
(WormBase), *Arabidopsis thaliana* (The Arabidopsis Information Resource;
TAIR), *Danio rerio* (Zebrafish Model Organism Database), and *Saccha-
romicae cervisiae* (Saccharomyces Genome Database) (see figure 1).[23] These
tools have played a particularly significant role in the development of online
data infrastructures in biology and continue to serve as reference points for
the construction of other databases to this day.[24] One reason for their popularity is their accessibility. Thanks to the sponsorship of national agencies,
they were made freely available for consultation by the whole biological
community, which enhanced their visibility as well as the status of model
organisms themselves as research objects.[25] Public funding also means that
their success was measured in terms of their popularity with biologists,
which pushed them to continuously improve in order to keep up with scientific demands. Another reason for the popularity is their focus on data documenting the *internal structure* of one species. Model organism databases
between 2000 and 2013 have not attempted to include data on organismal
behavior or on the environments in which specific variants typically grow in
the wild. In cases where the data are extracted from organisms collected in

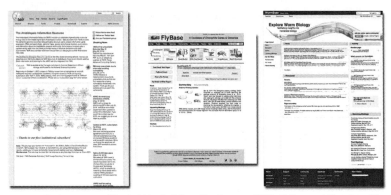

FIGURE 1 The homepages of The Arabidopsis Information Resource (TAIR), FlyBase, and WormBase. (The Arabidopsis Information Resource, https://www.arabidopsis.org; FlyBase, http://flybase.org; WormBase, http://www.wormbase.org/#01-23-6. All websites accessed June 2014.)

the wild, information about their geographical provenance and location has been included; but most of the data concern highly standardized organisms raised in laboratories under controlled conditions, thus effectively dismissing variables such as environmental variability and evolutionary time. This was not done out of indifference to the significance of these parameters but rather as a way to simplify the study of organisms to make it possible to integrate data about their composition, morphology, genetic regulation, and metabolism—an ambitious goal in itself and one that, as we will see, proved to be extremely challenging.[26]

TAIR, the main database collecting data on *Arabidopsis* up to 2013, is a good example of how central a role databases have played within the model organism research strategy. The National Science Foundation funded TAIR in 1999 to ensure that data coming out of the international sequencing project devoted to *Arabidopsis* would be adequately stored and made freely accessible to the plant science community. The Carnegie Institution for Science's Plant Biology Department, home to prominent plant scientists including *Arabidopsis* veterans Chris and Shauna Somerville, won a national bid to create and host the database. A former student of Chris Somerville, Seung Yon Rhee, was given the task of directing TAIR, which she undertook with a strong vision for what the database should become in the future. An experienced experimenter, Rhee thought that TAIR should not be just a repository for sequence data. Rather, it should become a repository for many data types generated by *Arabidopsis* research, a platform facilitating communication and exchange among researchers working on different aspects of plant biology. Furthermore, the database should contain a set of tools for data retrieval and data analysis, which would facilitate the integration of all those data; and it should enable users to compare *Arabidopsis* data with data extracted from other plant species, thus paving the way for developments in plant science as a whole.[27] Indeed, Rhee has been a staunch advocate of the value of sharing resources within biology, an approach she called "share and survive," as opposed to the "publish or perish" mentality characteristic of mainstream biomedical research.[28] In parallel to the success of *Arabidopsis* as a model organism for plant biology,[29] TAIR acquired increasing clout with researchers across the globe, assembling an impressive array of data sets concerning disparate aspects of *Arabidopsis* biology, ranging from morphology to metabolic pathways. By 2010, the database included various search and visualization tools elaborated by the TAIR team to help plant scientists in retrieving and interpreting *Arabidopsis* data. Examples are MapViewer, which allows access to various types of mappings of *Arabidopsis* chromosomes, and AraCyc, which visualizes data about biochemical pathways characterizing *Arabidopsis* cellular processes. TAIR provided abundant information about how these tools have been constructed, how they should be used, and which types of data are included.[30] It also cooperated with the *Arabidopsis* stock centers, which store hundreds of thousands of seed stocks of *Arabidopsis* mutants, so that users could order the specimens needed for their experiments directly from the website.[31]

The role of TAIR within plant science became a hot topic for scientific debate in 2008, when the National Science Foundation decided to cut funding to the resource. The motivations for this decision included financial constraints as well as the wish for TAIR to accommodate the changing needs of the plant science community, particularly the increasing importance of tools for the analysis and comparison of data across different plant species. At the same time, the decision to curtail TAIR funds, effectively making it impossible for its curators to extensively review and revise its contents, proved controversial. Scientists' protests poured in from all corners of the globe, *Nature* published an editorial on the key role of TAIR in plant science, and several working groups were set up to find ways to support TAIR in the long term, thanks also to the effort of the Multinational Arabidopsis Steering Committee (MASC) and the UK Genomic Arabidopsis Resource Network, or GARNet.[32] This ultimately resulted in the 2013 establishment of the Arabidopsis Information Portal (Araport), which is meant to complement and expand the original vision underlying TAIR by taking advantage of new and more sophisticated software, improved mechanisms for user feedback and data donation, and strong collaborative links to software developers and data storage facilities (such as those offered by iPlant) and cross-species efforts such as the General Model Organism Database (GMOD).[33]

FlyBase, WormBase, Mouse Genome Informatics, Saccharomyces Ge-

nome Database, and Zebrafish Database, like TAIR, were initially funded
by public agencies in order to store and circulate sequence data and yet
took advantage of the funding to increase the diversity of the data that they
host, thus functioning as sophisticated testing grounds for best practices
in data handling. This background hopefully clarifies why I chose model
organism databases as the main empirical terrain for my investigation of
data-intensive biology. These are exceptionally well-maintained, thought-
fully built resources whose goals and maintenance are strongly rooted in
the existing needs of research communities and whose importance to future
advancement is recognized and supported both by funders and by researchers. Because of their peculiar history, they are an excellent example of online
technology aimed at the dissemination of various types of data and constitute a good case study for analyzing what it means to care for data and how
such care affects both their production and their analysis.[34]

1.2 Packaging Data for Travel

Perhaps the main obstacle to the efficient dissemination of biological data
is the diversity of disciplinary approaches, methods, assumptions, and techniques characterizing biological research. This is evident in model organism biology, where each of the hundreds of research groups involved tends
to develop a specific epistemic culture, encompassing a unique ensemble
of skills, beliefs, interests, and preferred materials.[35] Moreover, biologists
typically adapt their methods and interests to the features of the organisms
that they study, thus further amplifying the already extensive differences
between research communities. This makes it hard to find ways to circulate data outside the situation in which they have been originally produced,
as researchers do not share a common terminology, conceptual apparatus,
methodology, or set of instruments. The global nature of biological research
makes travel even harder: data not only need to cross disciplinary and cultural boundaries, but they also need to travel great physical distances, becoming accessible to biologists regardless of their geographical location.

Much of the work involved in database development, particularly in
model organism databases, has focused on resolving the tension between
the local nature of data about organisms and the need for them to circulate
across widely different research contexts and locations. The work of the
people in charge of developing and maintaining databases, to whom I will
henceforth refer simply as "curators,"[36] is defined by the need to serve a
wide variety of database users across the globe, each looking for data suitable for their own interests and methods. Successful travel is marked by the
extent to which data are accessed and reused within new research contexts.

Notably, making data available online does not automatically make them usable. Whether data are fruitfully adopted across contexts is the result of packaging strategies developed by curators through years of specialized training and dialogue with users. These include the need to integrate different *data types* produced through various kinds of instruments and techniques, ranging from sequence data to photographs or tissue samples; to collect *metadata* that document the provenance of data (i.e., the conditions under which they were originally generated); to develop *representations* of data that facilitate searches and the visualization of results retrieved from databases (e.g., maps, models, simulations); to be able to order the *materials* on which data were originally acquired, such as specimens of the same mutant; and to adopt intelligible *keywords* for the classification and retrieval of data. In what follows, I examine these packaging strategies as providing the conditions under which model organism databases foster the travel of biological data across different scientific communities. As I will illustrate, "good packaging" consists of developing labels, infrastructures, and procedures that facilitate the retrieval and adoption of data by prospective users and is extremely hard to achieve in practice.

The process of packaging data for dissemination bears some similarities to the process of packaging items to be dispatched through the mail. The material tractability of traveling items is crucial in both cases: standard shape and dimensions help the packaging and circulation of the mail just as they help the packaging and circulation of data, and indeed, curators are often involved in long-term efforts to standardize data formats across biology as a whole.[37] Furthermore, data are objects whose ability to travel depends on infrastructure designed for this purpose, as well as interventions by people other than their senders and receivers. Human activities and physical environments are equally important to the travel of data. Post offices, trucks, drivers, mail carriers, and mail sorters play a similar role to databases and their curators. There would be no travel without the digital platform provided by databases and the work put in by curators to design and use them as a vehicle for data. And just as the mail is a service designed to satisfy senders and receivers, the need for data to travel is generated by the epistemic cultures in which data are produced and reused.

There are also important differences between packaging an object for express delivery and packaging data for dissemination. In both cases, whether travel is successful depends on whether what is packaged arrives at its destination without being damaged or lost, and which destination this will be depends on the way in which objects are labeled. However, in the case of data, labels *should not fully determine* the destinations to which the data will travel. There is no doubt that the labels chosen by curators have a strong influence on the direction that data will take. This is unavoidable, since the function of labels is precisely to make data retrievable by potential users. Without labels, data would not travel at all. Yet for successful reuse to take place, the journey that data ultimately undertake should be determined as much by their users as it is by their curators. Curators cannot possibly predict all the ways in which data might be used. This would involve familiarity with countless research programs, as well as a degree of scientific understanding and predictive ability that transcends the abilities of one individual or group. Therefore, the best way to explore and maximize the value of data as evidence is to enable as many researchers as possible to use data in their own way and within their own research context. Contrary to mail users, the prospective users of data circulated through databases are not passive recipients. For effective dissemination to take place, those users need to take an active part in the process of data retrieval, including a critical assessment of which labels to trust, which data to select, and how those data should be interpreted. Remarkably, curators of model organism databases are typically fully aware of the importance of representing multiple users' concerns within their practices.[38]

Given these premises, labeling becomes one of the most challenging components of the packaging process. Curators are required to create labels that, while making data retrievable by database users, do not prevent users from making their own selection of which data they wish to pick and how they want to interpret them. To pursue the analogy with mail packaging one last time, database labels need to indicate the information content of data without adding indications—such as a mailing address—about where the data could be delivered. Giving data the flexibility to travel wherever they might be needed constitutes a crucial characteristic of their packaging, which makes it much more sophisticated than the packaging of an object for travel to an already well-defined destination.

1.2.1 Relevance Labels: Bio-Ontologies. Enhancing data usability involves making data visible and accessible to as many researchers as possible. One labeling system in model organism biology has acquired enormous popularity precisely because it classifies data according to their relevance to investigating biological entities. This labeling system, known as "bio-ontologies," consists of a network of terms, each of which denotes a biological entity or process. Data are associated with one or more of these terms, depending on whether they are judged to be potentially relevant to future research on the entities to which the terms refer. For instance, gene VLN1 has been found to interact selectively with an actin filament known as F-actin.[39] This is an interesting finding given the crucial role played by the actin protein in several cellular processes, including motility and signaling. Nonetheless, the actual functions of VLN1 are still unknown: apart from its interaction with F-actin, there is no wealth of knowledge yet to associate with data about VLN1. Database curators tracked the available data about VLN1 and they classified them under the following terms: "actin filament binding," "actin filament bundle formation," "negative regulation of actin filament depoly-merization," and "actin cytoskeleton." Thanks to this classification, users interested in investigating these processes will be able to retrieve data about VLN1 and use them to advance their understanding.

Depending on which entities they aim to capture, there are many bio-ontologies in use in contemporary bioinformatics.[40] One of the most popular ones, from which I took the example above, is the Gene Ontology (GO), which encompasses three types of biological entities: cellular processes, molecular functions, and cellular components (Ashburner et al. 2000). Since their introduction in the late 1990s, bio-ontologies have come to play a prominent role in databases of all types, ranging from genetic databases used in basic model organism research to medical databases used in clinical practice.[41] One of the main reasons for this success is the way in which bio-ontology terms are chosen and used as labels for data classification.

Database curators select these labels according to two main criteria. The first is their intelligibility to practicing biologists, who need to use those labels as keywords in their data searches. In a bio-ontology, each biological entity or process currently under investigation is associated with one (and only one) term. This term is clearly defined so that researchers working in different areas can all understand what it is supposed to denote.[42] Often, however, different groups use different terms to formulate claims about the same phenomenon. This makes it difficult to agree on one term that could be used and understood by everyone interested in that entity—and in the data relevant to its study. Curators tackle this problem by creating a list of synonyms for their chosen label or by diversifying the labels themselves to reflect the terminologies used by different communities. The second criterion for the selection of labels is their association with datasets. The idea is to use only terms that can be associated with existing datasets: any other term, whether or not it is intelligible to bio-ontology users, does not need to be included as it does not help classify data. Curators create an association between a dataset and a term when they have grounds for assuming that the dataset provides information about the entity denoted by that term. This happens mainly through consultation of data repositories, where data are categorized as resulting from the experimental manipulation of the entity denoted by the term, and of publications using data as evidence to establish a claim about the entity denoted by the term.

Thanks to bio-ontologies, researchers can identify data stored in databases that might be relevant to their research interests. The focus on phenomena rather than methods or specific traditions makes it easier for researchers to bridge across the epistemic cultures in which data are originally produced. In this way, researchers with widely different backgrounds (in terms of methods and instruments used, discipline, or even theoretical perspective) can access the same pool of data and assess their relevance to their research. It therefore becomes more likely that the same data are used as evidence toward the validation of knowledge claims about the same entity. Thus labels such as bio-ontologies constitute a promising first step toward the packaging of data for successful reuse. They are not, however, sufficient for this purpose.

1.2.2 Reliability Labels: Metadata. The successful reuse of data depends also on the interpretation that researchers ultimately give them. For instance, consider the famous case of the DNA photographs produced by Rosalind Franklin in 1952 and examined by James Watson without Franklin's permission. The circumstances under which those photographs, particularly the much-discussed Photograph 51, were used as evidence for DNA structure are complex and include triangulation with other data and the different institutional and professional situations of Franklin and Watson. From Franklin's notebooks, it seems clear that when Watson and Francis Crick announced their discovery, she was weeks away from articulating the same interpretation of the data. Still, after Watson and Crick's announcement, she did not abandon attempts to use her data and ended up developing excellent work on viruses by using them to answer different questions.[43] This episode illustrates that there is not necessarily a single "right interpretation" of data. Interpretation depends on users' background and interests, which again highlights the need for curators to package data in ways that enable the emergence of local differences in interpretation.

The emergence of differences in interpretation, and thus the successful reuse of data, depends on users' awareness of the procedures through which data were originally produced. These procedures define several characteristics of data that are crucial in determining their quality and reliability, such as, for instance, their format, the actual organism used, the instrument(s) with which they were obtained, and the laboratory conditions at the time of production. These same elements might be completely irrelevant to assessing the quality of data produced through simulation, where the algorithms and parameters used for modeling are key factors. Hence, in order to reuse data found through a database, users need to be able to check, if they so wish, the conditions under which data have been obtained. This is why database curators devised metadata as a second type of label providing information about the provenance of data. An example of this is "evidence codes," which are meant to provide essential information about the procedures through which data are produced. They include categories for data derived from experimental research, as in IMP (Inferred from Mutant Phenotype), IGI (Inferred from Genetic Interaction), or IPI (Inferred from Physical Interaction); data derived from computational analysis, as in IEA (Inferred from Electronic Annotation) or ISS (Inferred from Sequence Similarity); and even information derived from informal communication with authors (TAS—Traceable Author Statement) and intervention by curators (IC—Inferred by Curator). Evidence codes are associated with each set of data that shares the same provenance. Once users have found data they are interested in, they can click on the related evidence code and start to uncover the procedures through which the data have been produced.

Developing this kind of labeling is a genealogical exercise in which curators investigate and reconstruct the sources and history of the data that they annotate. Metadata provide access to the qualifications that endow data with what Mary Morgan calls "character," thus making it possible for data to be adopted and reused across a variety of context.[44] Without metadata, researchers would not be able to judge the reliability of the data found online, which is a function of who produced them, for which reasons, and in which setting. This information provides grounds to trust data displayed in a database and to compare them with other data. For instance, knowing that two datasets have been obtained through similar methods from the same type of organism would enhance a user's willingness to treat them as compatible, while finding that one was obtained experimentally and the other from simulation would constitute a warning against such an assumption.

1.2.3 Decontextualization and Recontextualization. I have shown that what makes databases into good packages for data is the opportunity afforded to their users to evaluate the relevance and the reliability of the data in question. The two labeling systems enable users to disentangle the activity of searching and comparing data from the activity of assessing the reliability and significance of data. Thanks to bio-ontologies, researchers accessing a database can find out which existing datasets are potentially relevant to the study of the entities and processes in which they are interested. Once they have restricted their search in this way, they can use evidence codes to examine information about data production. This second type of labels enables them to assess the reliability of the data that they located through bio-ontologies, and eventually to discard data that are found wanting according to users' epistemic criteria. Remarkably, the consultation of evidence codes or other kinds of metadata is not aimed to reduce the existing gaps (if any) between the epistemic cultures of data producers and data users. Users get access to as accurate a report as possible about the conditions under which data were originally obtained. This does not necessarily mean that they need to know and think precisely what the producers know and think about those data. Users do not need to share producers' ways of reasoning and doing research. Rather, the consultation of metadata enables users to recognize disagreements with producers concerning what counts as suitable experimental conditions, to reflect on the significance of such disagreements, and to form their own opinions on the procedures used to obtain data. Any judgments on the reliability of data necessarily depends on the user's viewpoint, interests, and expertise—which is why curators choose labeling systems that, at least in principle, allow each user to form her own opinion.

Packaging data for dissemination via online databases necessarily involves two complementary moves. The first move, for which database curators are typically responsible, involves the *decontextualization* of data from their context of origin. The labeling of data through bio-ontologies ensures that they are at least temporarily decoupled from information about the local features of their production, which enables users to evaluate the potential relevance of those data to their research purposes without having to deal with an overwhelming amount of information. Decontextualization is a way for data to lose the significance attributed to them in their original research context: the whole point of decontextualization is to make data adaptable to new research settings, which is achieved by stripping them of as many qualifications as possible. When choosing and applying bio-ontology terms, database curators operate in ways similar to librarians when classifying books or archivists when classifying documents. Data are labeled so that users coming to the database can use those classificatory categories to search for a content-relevant item and adopt it for their own purposes.

Identifying which data to take from a database is a crucial step, and yet it does not help researchers to decide how to use those data. In other words, while helping to decontextualize data for circulation, bio-ontology labels do not help to recontextualize data for use in a new research setting. This *recontextualization* is the second move required for the successful packaging of data. It enables users to evaluate the potential meaning of data by assessing their provenance through the consultation of metadata. This is necessary to identify the value of data as evidence, thus helping to build an interpretation of their biological significance in a new research setting. In this sense, the process of recontextualization is reminiscent of work conducted by curators in a very different setting: museum exhibits, whose visitors can best form an opinion about the cultural significance of the objects in display when they are given information about the history of those objects and their creators. Arguably, the best exhibits provide such information while also encouraging visitors to form their own evaluation of the material on display. Similarly, the provision of metadata serves to steer researchers away from implausible interpretations and helps them to make an informed assessment that takes account of the full history of the data, rather than solely their appearance at the time of consultation.

By enabling users to access decontextualized data, databases provide the differential access needed to make data travel across contexts. By providing evidence codes, databases facilitate the recontextualization of data while at the same time making it possible for them to shift character and significance depending on their new location. This modality of data reuse is particularly important in model organism biology, where the same data might acquire entirely different interpretations when examined by biologists working on different species and/or dissimilar research cultures. Through their vision of recontextualization, curators are attempting to enable biologists to pick up new data without necessarily having much in common in terms of their goals and expertise. For instance, researchers investigating the regulatory functions of specific genes are using databases to check what data are available on their gene of interest, how those data were produced, and on which species. This enables them to compare what is known about the behavior of the gene across species, without having to become a specialist on each type of organism and experimental procedure involved.

The packaging strategies deployed by model organism databases function as a crucial material conduit and conceptual scaffold for data dissemination. Their physical characteristic and technological features, such as the option to layer information through clickable links and keep metadata separate from data themselves, are key to the coupling of de- and recontextualization involved in data travels. Paying attention to these processes shows that technology matters a great deal to the implementation of data-centric science. The opportunities opened up by computing tools have a significant impact on the ways in which researchers can retrieve, assess, and analyze data circulated online. At the same time, whether databases make data travel effectively depends just as strongly on the skill and ingenuity with which curators and users handle data retrieval processes.

1.3 The Emerging Power of Database Curators

The sheer number of researchers and the size of investments involved are not, by themselves, indicators of whether a database will be popular and sustainable in the long term.[45] No matter how complex and expensive, a data infrastructure remains worthless as long as there are no research groups adopting it and using it for their own purposes—and the more of those groups there are, the better. Within biology, this means that online databases are most successful when they mold themselves to the needs of multiple epistemic communities. Curators are well aware that the success of their products depends on how useful they prove to be to biologists, as this determines the levels of funding and community support that they will receive. Their careers depend at least in part on their ability to identify, embrace, and constructively engage with as many epistemic cultures in biology as possible. This means making their choice of labels at least compatible with, and at best conducive to, widely diverse forms of computational modeling and physical intervention on actual organisms. Ideally, users should be able to interpret data in ways that depend solely on their own backgrounds and interests. In practice, however, curators are responsible for identifying what additional information is needed to recontextualize data into new research settings. In other words, they are in charge of decontextualizing data, while also making sure that users can access whatever metadata they need to be able to evaluate and interpret data.

Balancing these two requirements against each other is not easy. Nor, given the ever-changing nature of the data and practices involved, are there universal and enduring ways to coordinate decontextualization and recontextualization strategies. Database management is a dynamic process whose functioning depends on the degree to which curators manage to capture the changing wishes and constraints of practicing biologists. Within model organism databases, this involved a great deal of judgment exercised by curators at all stages of the process, starting from the very gathering of data for inclusion in the system. Particularly at the start of these projects, data collection involved the extraction of data from existing publications and repositories, which forced curators to single out publications that they considered to be reliable, updated, and representative for specific datasets. For instance, TAIR curators wishing to collect data on a given gene (say the Unknown Flowering Object [UFO] gene in *Arabidopsis thaliana*) could not compile data from each relevant publication, as it would be too time consuming: even just a keyword search on PubMed for "UFO Arabidopsis" results in over fifty journal articles, only one or two of which are used as reference for an annotation. Hence curators chose what they saw as the most up-to-date and accurate publications, which as a consequence became "representative" publications for that entity. Once this was settled, curators also had to assess which data therein contained should be extracted and/or how the interpretation given within the paper matched the terms and definitions already contained in the bio-ontology. Does the content of the paper warrant the classification of given data under a new bio-ontology term? Or can the contents of the publication be associated to one or more existing terms? These choices are impossible to regulate through fixed and objective standards. Indeed, bioinformaticians have been trying to automate the process of extraction for years, with little success. The very reasons why the process of extraction requires manual curation are the reasons why it is hard to divorce it from subjective judgment: the choices involved are informed by a curator's expertise and her ability to bridge between the original context of data production and that of data dissemination.

Performing curation tasks such as extraction presupposes skills honed through specific training and years of experience. The best curators are veritable "packaging experts." They need to have some familiarity with various fields of biological research, which helps them to recognize and respect the diversity characterizing different epistemic cultures and associated terminologies, norms, and methods. At the same time, they need to couple a generalist understanding of biology with an awareness of how research "at the bench" is conducted, so as to be able to identify the parameters that users would like to see as metadata. Most curators working on model organism databases and the Gene Ontology in its first decade of activities were biologists by training and motivation. Many of them had trained in at least two different biological subfields before moving into curation, and their decision to extend their expertise toward computer science and bioinformatics was primarily due to their interest in improving data analysis tools for model organism research as a whole.[46] The curators' hands-on knowledge of experimental work was reflected in the development of databases and associated labels that would be intelligible to experimenters.

Of course, curators also have to have a good understanding of cutting-edge information technology, so as to be able to collaborate with programmers and computer engineers in developing appropriate software. This is a complex requirement due to ongoing tensions between computer scientists and biologists concerning the criteria and priorities to be adopted in bioinformatics and not least to the differences in training and objectives characterizing these two fields.[47] Within model organism databases, the priority was clearly given to biologists, with the idea that they are the ultimate users of the tool being produced. Biologists needed however to be pushed to recognize and adopt some of the solutions—and associated constraints— provided by computing, which often challenge well-entrenched ways of thinking about data dissemination and analysis. Even the simple requirement that all data inserted in the system had to be machine readable became ground for serious disagreements, given the abundance of complex images (from microscopy and mass spectrometry, for instance) that biologists consider as crucial data for their work. Such data continue to be difficult to analyze computationally and require complex annotations including the labeling systems mentioned above.

Beyond the problems involved in bringing together IT and biology, there are also tensions involved in attempting to create labels that cut across the difference in terminologies, methods, and interests characterizing biological subdisciplines. This affects both relevance and reliability labels, since existing differences in how investigators record and classify information about data production and interpretation give rise to what Paul Edwards, Geoffrey Bowker, Christine Borgman, and their collaborators have aptly called "data frictions."[48] I return to those frictions, and what they involve in the case of biology, in chapters 4 (metadata) and 5 (bio-ontologies). For the time being, all I want to stress is the role of curators as unique witnesses of, and key mediators between, the diverse concerns, interests, and expertises involved in making and using an online database. By taking upon themselves the task of choosing the appropriate package for data, curators unavoidably make important decisions on what counts as relevant data for any specific research project. Users could play a significant role in shaping these choices, but in practice, most of them are happy to trust curators with this role, as they do not want to spare time and energy from their research to deal with choices about data packaging. For this same reason, however, users are reluctant to invest effort in understanding the choices made by curators. Users want an efficient service by which they can access a database, type a keyword, get the relevant data, and go back to their research. By thus relegating packaging responsibility to curators, they often do not understand the extent to which packaging affects the travel of data and the ways in which they will be reused.

Curators are well aware that their interventions influence where and how data will travel. To some extent, they endorse what might be called a "service" ethos: they are willing to recognize that it is their professional duty to serve the user community as best as they can, and they feel both responsible and accountable for their packaging choices. The language of "service" is crucial to making databases attractive to prospective users and is often used by curators to describe their work to biologists (as in the case of iPlant). Yet the description of curation as a "service" constitutes an obstacle to its recognition as an important part of science in its own right. The idea that their work is meant to facilitate experimental research can be taken to imply that their contribution is perceived as "second class," rather than as complementary to it. In the current scientific credit attribution system, services such as bioinformatics and database building are typically recognized not as forms of research but rather as technical means to the conduct of research, in the same category as scientific instruments and material infrastructures such as laboratory space. To counter this perception, curators are actively seeking scientific recognition for their role as packaging experts, and biocurator meetings are now regularly held to facilitate cooperation and interoperability among databases across the globe.[49]

Curators are aware that it is impossible to conform to the expectations and practices of rapidly changing fields without being in constant dialogue with the relevant user communities. This is also because, aside from one-to-one dialogue and website statistics on which parts of a database are most popular with users, there is currently no reliable way for curators to systematically evaluate how users are employing information in the database. Many researchers are not yet used to citing databases in their final publications—they would rather cite the papers written by the original producers of the data, even if they would have not been able to find those papers and associated data without consulting a database. It is therefore often difficult for curators to assess which research projects have made successful use of their resources. Nevertheless, many attempts to elicit feedback fail because of users' disinterest in packaging practices and their inability to understand their complex functioning. The gulf between the activities and expertises of curators and users tends to create a problematic system of division of labor. On the one hand, curators invite users to critically assess their work and complain about what they might perceive as "bad choices." On the other hand, users perceive curators' work as a service whose efficiency should be tested and guaranteed by service providers rather than the users. They thus tend to trust curators unconditionally or, in the absence of trust, simply refuse to use the service.

These tensions are exemplified by a recent attempt to disseminate data about leaves within plant science, in which I was briefly involved as an observer. AGRON-OMICS was a European project sponsored by the Sixth Framework program between 2006 to 2010 that brought together plant scientists from a variety of laboratories and disciplines, including molecular, cellular, and developmental biology. Its goal was to secure an integrated understanding of leaf development by gathering and analyzing data extracted from the model organism *Arabidopsis thaliana*. A crucial component of this project was the search for efficient tools to circulate data among members of the group and to the research community at large. The question of labeling was thus uppermost in the minds of the group coordinators from the outset in 2006. What categories could be used to circulate data gathered by researchers so steeped in their own local terminologies and practice? The very first meeting of the project, a two-day workshop titled "Ontologies, Standards and Best Practice" was devoted to tackling this question.[50]

Participants included the main scientific contributors to AGRON-OMICS and the curators of the databases that were most likely to be of use, such as Genevestigator, the Arabidopsis Reactome, the Gene Ontology, and the Plant Ontology. Curators did most of the talking, both through presentations explaining what their tools could do and through hands-on workshops teaching researchers how to use them. Most questions raised concerned systems for tracking the relevance and reliability of data; users and curators agreed on the importance of keeping the focus on these two factors. Overall, the workshop was successful in alerting researchers to the importance of finding good packages to make their data travel. Remarkably however, this lesson came with an increased awareness of the difficulties plaguing these efforts, particularly of the problems associated with labeling data for reuse.

Many of the scientists in attendance displayed distrust for the work of curators, which they saw as far removed from actual biological research. The very need to decontextualize data was seen as potentially problematic, despite evidence for the necessity of this process to make data travel. There were complaints that curators, in their tight collaboration with computer scientists, tended to favor a polished labeling system over one that would actually help experimenters. It was also remarked that the synonyms system devised by curators to accommodate terminological pluralism only worked if curators were aware of all existing synonyms for a given label. Further, some researchers were dazzled by the multitude of tools available for labeling (well over twenty were mentioned at the meeting, most of which researchers were not yet acquainted with; and labeling systems have continued to proliferate since). While some labels, such as the Gene Ontology, are fairly well established across a number of databases, there are many cases of databases developing their own labeling systems without regard for the ones already in place. This leads to a proliferation of labels that is confusing to most users, who feel they are wasting time in learning to use all those systems and in assessing each label's merits relative to others. Although some scientists appreciate the idea of being able to choose among different labeling tools, this is often associated with an interest in developing those tools themselves.

Dialogue between users and curators over these difficulties resulted in both sides increasing their understanding of labeling processes. Curators walked away with a better idea of the needs and expectations of AGRON-OMICS researchers. Users however retained a high degree of skepticism toward curators' work. Indeed, precisely as they were learning to appreciate the scope and implications of curators' work, AGRON-OMICS scientists saw the importance of selecting appropriate labels for their data, as well as the power that this brings over the eventual reuse of those same data. They therefore resolved to take over some of that work in order to ensure that the labels used to package data be perfectly suited to their research needs. One of the action points agreed on at the end of the meeting was the creation of two new bio-ontologies: one for *Arabidopsis* phenotypes and one for *Arabidopsis* genotypes. The main rationale for this effort was the perceived absence of suitable labels dealing with these biological entities. Also, developing their own labels would ensure that scientists take over the packaging of data of particular importance to their project. This sound reasoning did not, however, take due account of the complex and time-consuming technical aspects involved in implementing such new systems. The project succeeded in developing and maintaining its own database on *Arabidopsis* phenotypes, Phenopsis DB, which supported new findings on growth-stage analysis and was still functioning at the time of writing; but it did not manage to set up a genotype database, ultimately preferring to rely on other existing databases despite their less-than-ideal fit.

The AGRON-OMICS case shows how, paradoxically, the very expertise that enables curators to develop and maintain a database constitutes an obstacle to the communication between curators and database users. Many researchers do not have the skills to provide feedback to curators on how well their systems serve their research. Providing feedback unavoidably means engaging with the practices through which bio-ontologies are developed, and thus acquiring some of the skills involved in curation. Understandably, given the time, interest and effort involved, this is something that database users are often reluctant to do. A molecular biologist I interviewed in March 2007 summarized the problem as follows: "Biologists just want to get information and then go back to their question." To researchers subscribing to this view, the elaboration of packaging strategies is not a matter of democratic consultation over which terms and definitions to adopt, but rather a matter of division of labor between people busy with experiments and people busy with developing databases storing the results of experiments. In their eyes, the production of a reliable labeling system is the job of curators; all they should need to do is trust the curators' judgment.

AGRON-OMICS is just one of many examples illustrating how the integrative efforts of curators are not well coordinated with the data integration carried out by database users, such as experimental biologists. Scientists involved in generating knowledge through data reuse do not typically want to spend time participating in curatorial efforts, and the lack of recognition for curatorial activities within systems of credit attribution only adds to this problem. Even more problematically from the epistemic viewpoint, many biologists do not fully appreciate that data dissemination via databases involves substantial integrative efforts, rather than being simply conducive to making data travel. This is partly due to the idea that data found online are "raw"—that is, that they are shown online in exactly the same format as when they were first produced. Biologists who are committed to this idea are reluctant to consider how the integration of data affected through databases is likely to affect the original format of data and the ways in which they are visualized. So databases are envisaged by many scientists as a neutral territory through which data travel without changing in any way; while, in order to work efficiently in disseminating data, databases need to function as a transformative platform, within which data are carefully selected, formatted, classified, and integrated in order to be retrieved and used by the scientists who may need them.[51]

Curators are not simply responsible for making data travel; they are responsible for making data travel *well*, which involves communicating with users to make sure that data are indeed being reused. In the case of AGRON-OMICS, the potential tensions between curators and users were resolved by making these two figures overlap. It is not clear, however, whether this is a good solution. In the absence of a generalist curator aiming to serve the whole biological community, the labels used for packaging might end up serving the needs of the AGRON-OMICS group over and above the needs of other scientists, hampering the successful reuse of those same data in other quarters. Furthermore, as I already mentioned, few scientists are willing to invest time and effort toward the creation of good packages for data. Part of AGRON-OMICS funding was explicitly directed to the study and testing of packaging tools for data, which meant that they could employ people to work on bioinformatics and they had resources for developing and maintaining communication with curators at the international level (preventing the danger of narrowing their vision to their own project). The same is not true of smaller projects with more specific goals. A more general solution is to enforce some mechanisms of communication between curators and users so that curators receive frequent feedback from the widest range of users, ensuring that their packaging strategies are indeed serving the needs of users as they evolve through time. In other words, packaging—and particularly the decontextualization processes for which curators are responsible—requires external regulation, a topic that I explore in more detail in the next chapter.

1.4 Data Journeys and Other Metaphors of Travel

I have examined the development and features of online databases in biology, paying specific attention to their adoption within model organism research and the ways in which they have come to serve as key infrastructures in this area and beyond. I stressed how computer technologies and the Internet are offering opportunities to realize long-held aspirations within biological communities, such as the ability to disseminate information. I also pointed to the difficulties involved in balancing the computational skills needed to develop functioning databases with the vast diversity of biological expertise needed to package data for prospective recontextualization. Readers will hardly have failed to notice that I framed my whole analysis in terms of data movement and travel. This is a deliberate choice, which I wish to defend explicitly in this final section. Indeed, I want to suggest that *data journeys*, which designate the *movement of scientific data from their production site to many other sites within or beyond the same field of research*, are a defining feature of the epistemology of data-centric biology, which marks both its relative novelty as a historical phenomenon and its peculiarity as an approach to scientific inquiry.

Data-centric science is inextricably tied to the widespread dissemination of large quantities of data. Such big data are by no means a new phenomenon in the history of biology, and biologists have been attempting to improve procedures for the preservation and circulation of data since at least the early modern period, as illustrated by the complex displacement strategies involved in assembling materials for large natural history collections.[52] However, the development of computational methods, technologies, and infrastructures has opened up opportunities for data dissemination and made it possible to pursue research goals that biologists in earlier periods had sometimes dreamt about but found impossible to implement. For example, I have mentioned the efforts put by *Drosophila* researchers in the early twentieth century, particularly T. H. Morgan's group, to make their data and specimens widely accessible to as many interested biologists as possible. These efforts were frustrated by the need to make dissemination procedures into personal transactions. The main vehicles for data were international newsletters, which could only contain small amounts of information; as a result, individuals interested in reusing some of Morgan's data would need to approach his group and acquire additional information directly through communication with them.[53] This system could not function in the kind of large, global research environment in which most biologists work today, where making personal contact with data producers is often too time-consuming to be feasible, particularly when attempting to scope the field for exploratory purposes. Online digital technologies make it possible to disseminate information in a highly automated, impersonal way, while also fostering a global reach and the opportunity to modify and update the system instantaneously. Given these advantages, it is hard not to focus on data movements in space and time as a key feature of data-centric research.

As my analysis has hopefully made clear, however, there are serious problems with the promise of smooth and widespread movement that is often associated with big data. This is where looking at strategies for data packaging and dissemination, while at the same time evaluating their relation to contexts of reuse in biology, is crucial to understanding the dynamics of data-centric science. Thinking about data journeys is important because journeys are hardly ever unproblematic. Journeys require long-term planning, reliable infrastructures, and adequate vehicles and demand energy and work, as well as a considerable amount of financial resources. They may be short or long, fast or slow. They can happen in a variety of ways and for a variety of reasons. Often they require frequent changes of vehicles and terrains, which in turn force travelers to change their ways and appearance to adapt to different landscapes and climates. Furthermore, journeys can be interrupted, disrupted, and modified as they unfold. Travelers may encounter obstacles, delays, dead ends, and unexpected shortcuts, which in turn shift the timescales, directions, and destinations of travel. What was planned as one short-term journey may end up involving several trips over a long period of time, while what was envisaged as an extensive trip may be cut short by changes in goals, personal circumstances, or lack of fuel. These qualities are the reasons why I view the metaphor of travel as useful when discussing data movements, as also exemplified by Bruno Latour's work on chains of reference and the mobility of evidence across extended networks of actors and Mary Morgan's analysis of the importance of vehicles and companionship in the travel of facts.[54]

Bringing the metaphor of travel to bear on data-centric biology highlights that there is nothing smooth about data journeys. Like human journeys, they are typically complex and fragmented, often involving planning ahead of time and resorting to several types of media, social interactions, and material infrastructures—in other words, a lot of labor and investments, as I will explore in the next chapter. They may range from very concrete shifts of materials from one individual to another (as when a researcher shows her latest data to a colleague in the same lab without giving them a copy, thus retaining control over who gets to see the data and how) to highly diffused and depersonalized dissemination such as the ones mediated by online databases, where the whole point of the system is to push data in unpredictable directions, thereby losing control over where data may end up (a feature which, incidentally, makes it hard if not impossible for data journeys to be planned over long time scales—the future of communication technologies and databases themselves being hard to foresee beyond the next twenty years[55]). Additionally, data that go through journeys are rarely unaffected. Travel can affect their format (e.g., from analog to digital), their appearance (when they are visualized through specific modeling programs), and their significance (when they change labels, for instance when entering a database). Travel also typically introduces errors, such as those caused by technical glitches plaguing the transfer of data from one type of software to another, power cuts or lack of storage space when moving data to a new server, or typos made when manually inserting variables into a database. These observations mark an important difference between my account and Latour's emphasis on the immutability of data as the source of their mobility. In my analysis, data journeys depend on the mutability of data and their capacity to adapt to different landscapes and enter unforeseen spaces; by the same token, they are affected by the unavoidable serendipity involved in any type of displacement.

These features are particularly important to keep in mind, given the flurry of water-related metaphors often associated with data-centric science in the popular and scientific press. Ideas such as "data flood," "data deluge," and, most recently, "data flows" seem to suggest that the dissemination and reinterpretation of big data is a fluid and unproblematic process—a process wherein as long as data are somehow circulated, they will be magically transformed into new knowledge. The rest of this book is devoted to analyzing the challenges, achievements, and obstacles involved in disseminating data to enhance the production of scientific knowledge. Not only do data not "flow" toward discovery, but it is the lack of smoothness and predefined direction that makes their travel epistemologically interesting and useful. Furthermore, the idea of data flow seems to suggest that data travel as a cohesive ensemble, which like a river moves effortlessly and in compact ways from one place to another.[56] I agree that data tend to travel as a group (or "set"), but the composition of the group can also vary greatly as travel progresses. In shifting from laboratory to publication, publication to database, and database to new research environment, data can be lost, acquired, misrepresented, transformed, and integrated—and the metaphor of a journey seems to better capture these features of mass movement than the notion of flow.

Another term that I find problematic, and yet is often used to refer to data movements, is the idea of "data sharing," which unavoidably evokes ideals of reciprocity and community building by depicting data dissemination as a form of exchange between individuals and/or groups. This seems to suggest that practices such as data donation (the deliberate decision, on the part of an investigator, to release data to a public repository, database, or publication) are somehow reciprocal, involving an exchange between two well-identified parties. This may well be the case in some situations, and the idea of reciprocity is certainly widely exploited in the collection of medical data from patients, where arguments about "giving something back" in exchange for treatment are often used to convince individuals to donate their personal data to scientific research,[57]; as well as in model organism communities, where some curators argue that whoever benefits from access to a database should also contribute to its development, for instance by donating data. However, reciprocity is by no means the norm when considering all forms of data dissemination practices in biology. The example of model organism databases is itself a demonstration of the fundamental lack of reciprocity in current practices of data dissemination. These databases are publicly accessible, so anyone can access them regardless of whether they have contributed data to it. However, donating data is not typically perceived by researchers as a rewarding experience, both because of investigators' feeling of ownership over their data and because the scientific credit regime does not value this kind of activity. As a result, curators struggle to attract data donors, and donations become even more elusive in highly competitive fields. For these reasons, I do not like to use the term "sharing" to depict the practice of data dissemination.

I should point out that the notion of data journey is not itself devoid of problematic associations. For instance, the idea of journey can be rightly critiqued as overly anthropocentric for entities such as scientific data—after all, human journeys are enacted and performed by individuals who have their own desires and volition and who take decisions over what to do next. Within my account, I do not wish to attribute agency nor intentionality to scientific data, though I will point to some of their material characteristics (such as their format, features, and size) as important factors in determining their journeys. Data have no agency in the sense that they do not themselves perform work in order to travel. What makes them move is externally generated and involves many more interventions than the technical ones examined in this chapter. Another set of concerns is tied to the potential interpretation of journeys as themselves linear enterprises, which proceed orderly from one point to another and where what happens along the way has no repercussions on previous stages of travel. It should be clear by now that I do not wish to convey this impression, and my use of the metaphor of the journey is rather intended to complement ongoing work in the history of science around the challenges, dynamics, and opportunities involved in moving knowledge around (whatever one takes "knowledge" to be at any point in time).[58] Particularly notable for my purposes is the ongoing debate concerning the very terms "dissemination" and "circulation," which I have hitherto used without critical qualifications. Historians such as Sujit Sivasundaram, Kapil Raj, Simon Shaffer, and Lissa Roberts have explored the risks involved in the unqualified adoption of these notions, which include overlooking the heavy asymmetries and complex translations involved in movement, exchange, and brokerage and the ramifications that any one event or intervention may have on the journey as a whole.[59] These are substantial worries, and it is no surprise that they should arise from scholarship focused on colonial and postcolonial histories, where tracing who provides what information, when, and how is crucial to overcoming traditionally Eurocentric assessments of what scientific and technological knowledge consists of and who participates in its production. In the same spirit, my use of the notion of data journey aims to highlight the enormous amounts of work and ingenuity involved in developing material, conceptual, social, and institutional means to package data for travel while also stressing the disparities and divisions involved in determining how packaging can work, what it means to "travel well," and for whom. This will hopefully discourage readers from assuming that the dissemination of data, or any other components of research, can entail anything less than engagement with the full spectrum of activities, materials, and institutions supporting data handling (whether or not such engagement is acknowledged or even apparent to the individuals involved).

Emphasizing the role of databases in managing and fostering epistemic diversity in science calls into question the rhetoric of data reuse employed by funders and database curators alike. Facilitating data reuse is extremely complex, requiring multiple displacements: research moves from existing research projects to databases to new projects, with high potential for misunderstandings across different research loci. Acknowledging this leads to the recognition that making data accessible is a different and arguably less complex challenge from making them reusable. This apparently simple point keeps being overlooked in current governance of large-scale biology, where expectations about data reuse are typically not backed up by extensive empirical studies of the needs and expectations of actual users. Curators attempt to interpret those wishes as well as they can, and these efforts need to be recognized as crucial to the success of large-scale databases. However, the fact that no systematic and empirically grounded research grounds these intuitions is striking and troubling given the level of financial and scientific investments in these resources.

The hype attached to database development as an easy solution to the "data deluge" has taken attention away from the problems involved in actually using data found online toward further research: in particular, from the difficulties of matching in silico representations of the world with experimentation in vivo and clinical intervention and also aligning the experimental practices characterizing research on humans with the ones used to research model organisms. I have analyzed here the processes of database development not primarily as a means toward the solution of those problems (thought this might certainly be the case) but rather as a site where diverging stakes, values, and methods characterizing research cultures in biology can be identified and discussed.

2 Managing Data Journeys: Social Structures

In the previous chapter, I stressed the scale and sophistication of the technical skills and material resources required to make data travel and concluded that data packaging is far from being a purely technoscientific problem. Whether data are disseminated, to whom, and to which effect depends on more than the labels, software, and norms promoted by database curators. Data infrastructures are costly, involving extensive, long-term investments in both human and material resources. Furthermore, the fact that data are produced by a variety of groups, for different purposes, in different parts of the world poses immense logistical, political, ethical, and structural challenges to their dissemination, which in turn affect the ways in which data are circulated and interpreted by biologists. Hence whether data are disseminated, to whom, and to which effect depends on the existence of relevant regulatory and social structures, including institutions and networks of individuals that take responsibility for developing, financing, and overseeing data infrastructures vis-à-vis complex and ever-changing political scenarios and economic demands, and make significant choices concerning the present and future value of the data themselves.[1] In Mike Fortun's apposite terminology, data need to be *cared for* in order to yield knowledge; and whoever provides such stewardship, whether it is data users, curators, funding bodies, or scientific institutions, is bound to regard data as forms of scientific, social, and/or economic capital.[2]

These observations have profound implications for how scientific data, and their dissemination, are conceptualized and studied. They highlight the significant role played by collectives in promoting and structuring data-centric research, so as to maximize their value as evidence for discoveries to come. They also underscore how the value of scientific data goes beyond the evidential, imbued as it is with financial, political, cultural, and even affective dimensions, and how this multidimensionality nurtures and structures their use within research. In this chapter, I examine the multiple ways in which data are valued, and the epistemological implications of this situation, by considering how data packaging has been institutionalized within biology and its relation to the political economies of globalized research. This is crucial to understanding the prominence of data centrism today and the difference between this development and previous moments in the history of science where data attracted public and scientific attention.

2.1 The Institutionalization of Data Packaging

As exemplified by the dispute that erupted in the early 1990s over how data obtained through the Human Genome Project should be released, data dissemination in biology is a controversial matter, not least because it highlights the interdependencies between the norms and practices sanctioned within research communities and the broader institutional, economic, and cultural regimes within which these communities operate.[3] The factors involved in regulating the disclosure and circulation of biological data range from the conflicting interests and ethos of the researchers involved to the clash in goals and procedures between biotechnology and pharmaceutical industries, national governments, and international agencies. Given the extent of investment involved in contemporary biology and biomedicine, questions about ownership and authorship of data, and the accountabilities that such claims involve, have become ever more significant. Data producers worry about the loss of control that may come from making their results widely available. Part of this worry is linked to fears of being "scooped" by other laboratories interested in the same topics, but there are also significant concerns around the possible misuse of data taken out of context and confusion around which regimes of intellectual property pertain to the results of academic research (particularly when such research is partly or fully funded by private companies).[4] Scientific institutions and funding bodies are attempting to ward off these fears by adding credibility as well as accountability to activities of data production. A key example of this is the increasing recognition of data authorship, certified by publication in data journals and citable databases, as part of the metrics used to evaluate the quality of research. This is expected to act as an incentive for scientists to freely disseminate data, thus increasing the number of datasets that are publicly accessible and enabling data authors to acknowledge responsibility if data turn out to be "bad" in some way or are mislabeled in the process of dissemination.[5] At the same time, data publication practices raise the question of how to value the input of researchers, such as database curators, who are involved in the formatting, classification, dissemination, and visualization of data (in other words, those involved in facilitating data journeys beyond the stage of data production). Whether these people should also be seen as "authors" and/or "owners" of data, or instead "contributors" to data use, and what this means in terms of knowledge production and institutionalization are among the most difficult issues confronted by science policy bodies at the time of writing.[6]

This situation of uncertainty around the status and nature of data handling practices has given rise to new types of organizations, often created by scientists themselves, aiming to function as platforms for networking, debate, and joint action around research management. Many such organizations in biology call themselves "consortia," thus stressing their commitment to tackle a common set of concerns, which can span from interest in a specific phenomenon (e.g., the Beta Cell Biology Consortium, devoted to pancreatic islet development and function; http://www.betacell.org) to willingness to solve a common technical problem (e.g., the Flowers Consortium in the United Kingdom, aimed at creating a common infrastructure for synthetic biology; http://www.synbiuk.org) or to promote a specific standard or technique (e.g., the Molecular Biology Consortium [MBC], founded to further high-throughput analysis of biomolecular and subcellular structures via a superbend X-ray beamline at the Advanced Light Source; http://www.mbc-als.org). The members of a consortium, which can be individuals as well as groups, labs, and institutes, do not need to be located in the same geographical site or to belong to the same discipline. Indeed, the term is often used to designate groups of scientists based in different institutions around the world and coming from a variety of disciplinary backgrounds. Consortia are sometimes fueled by dedicated funding, most often provided by governmental bodies interested in supporting a specific area of scientific work; in other cases, financial support is achieved by bringing together a variety of resources on an ongoing basis, thus underscoring the bottom-up, collective nature of the institution.

In this section, I focus on a prominent example of the latter kind, in which biologists have created an organization specifically aiming to help in negotiating common standards and procedures for data dissemination. This is the case of bio-ontology consortia, committees of researchers that emerged in the early 1990s to develop and maintain the relevance labels employed by the model organism databases and to function as a much-needed interface between bottom-up regulation arising from scientific practice and top-down regulation produced by governmental and international agencies. Bio-ontology consortia achieve this by focusing on practical problems encountered by researchers who use data packaging tools such as databases. A good example is the problem of data classification—that is, the tension that is bound to exist between the stability and homogeny imposed by classificatory categories used in databases and the dynamism and diversity characterizing the scientific practices through which data are produced. Bio-ontology consortia provide an institutional solution to this problem by setting up mechanisms to select and update the relevance labels given to data in ways that mirror the expectations and needs of data users. These consortia are typically born out of the initiative of database curators, who are well aware of the problems surrounding data classification and labeling within data infrastructures, and decide to join forces in order to bring visibility to those issues within the wider scientific community. Over and above more traditional scientific institutions and funding bodies, it is these organizations that have taken responsibility for facilitating collaboration and dialogue among curators as well as between curators and users—and that therefore play a crucial role in the regulation of data sharing. Examining the circumstances in which these organizations emerge, as well as their effects on research practices and regulatory structures, illuminates aspects of governance in contemporary biomedical research and its impact on knowledge production. This is a case where the space of scientific governance is being reconfigured to facilitate the adoption of technologies for the production and exchange of data and to improve the ways in which they are used to generate scientific knowledge.[7]

An exemplary case is the Gene Ontology (GO) Consortium, which was instrumental in the development and current success of the GO as a classification tool. GO Consortium started as an informal network of collaboration among the curators of prominent model organism databases such as FlyBase, Mouse Genome Informatics, and Saccharomyces Genome Database.[8] These researchers were aware that the problems emerging in relation to data packaging could not be solved by individual initiatives. They therefore decided to use some of the funding allocated to each of their databases to support an international collaboration among database developers, which was aimed at the development of adequate labels for data and which they named GO Consortium. Institutionalizing their informal network into an independent organization served several purposes: it allowed them to attract funding specifically supporting this initiative; it gave visibility to their efforts among user communities, particularly since the consortium, rather than specific curators, was listed as the author of publications devoted to GO; and it gave other curators the opportunity to join in. Within less than a decade, the GO Consortium was able to attract funding from both private and public agencies (including a pump-priming grant by AstraZeneca in 1999 and a grant by the National Institutes of Health in 2000, which was then renewed), enabling the consortium to fund the employment of four full-time curators to work in the their main office in Cambridge, United Kingdom. At the same time, the consortium expanded to incorporate several new members, including most model organism databases.[9] At the time of writing, the consortium included over thirty members, each of which is required to "show a significant and ongoing commitment to the utilization and further development of the Gene Ontology."[10] This means funding some of their staff to work on GO and contribute to its content, sending at least one representative to GO Consortium meetings, and being prepared to host those meetings at their own institutions.

Another example is provided by the Open Biomedical Ontology (OBO) consortium, an umbrella body for curators involved in the development of bio-ontologies that was started by Michael Ashburner and Suzanna Lewis in 2001. The initial motivation was to develop criteria through which the quality and efficiency of bio-ontologies as classificatory tools could be assessed and improved. These included open access to data (with exceptions in the case of sensitive data such as those derived from clinical trials); active management, meaning that curators would be constantly engaged in improving and updating their resource; a well-defined focus, which would prevent redundancy between ontologies; and maximal exposure to critique, for instance through frequent publication in major biology journals (thus advertising the ontology and attracting feedback from potential users) and the establishment of mechanisms to elicit comments from users. Not incidentally, these have also been singled out as "the key principles underlying the success of the GO."[11] In other words, the OBO consortium set out to make the GO into an exemplar in the Kuhnian sense: a textbook example of what a bio-ontology should be and how it should function, as well as a "model of good practice."[12] At the same time, the OBO consortium used the feedback gathered through interaction between curators to develop rules and principles that could be effectively applied to ontologies aimed at different types of datasets.[13] GO itself ended up being substantially reformed as a result of this process. Within six years of its inception, over sixty ontologies had become associated with the OBO consortium (where association involves similar requirements for collaboration as membership in the

GO Consortium) and many more curators learned from the experiences gained through these cooperations. Participating ontologies range from the Foundational Model of Anatomy to the Cell Ontology, the Plant Ontology, and the Ontology for Clinical Investigations. In several cases, each of the participating ontologies also maintains its own consortium (for instance, the Plant Ontology Consortium), which again helps curators interact with experts in the specific fields addressed by the bio-ontology. Governmental agencies, again most notably the National Institutes of Health, have begun paying close attention to the efficiency with which consortia operate and rewarding them by allocating apposite funding.

The main function of bio-ontology consortia like the OBO and the GO is to effectively coordinate and enforce cooperation across three groups involved in the regulation of data dissemination.[14] The first group comprises *database curators*. Consortia provide an institutional incentive for the exchange of ideas, experiences, and feedback among curators busy with different projects, thus speeding up developments in bio-ontology curation, enhancing curators' accountability to their peers, increasing effective division of labor among curators, and at the same time helping maintain and legitimize a collaborative ethos. Exchanges are achieved through regular face-to-face meetings and weekly communications through various channels, ranging from old-fashioned e-mails to wikis, blogs, and websites (such as the BioCurator Forum[15]). Indeed, consortia play a significant role in training curators, during a time in which the professional profile of these figures is far from established, and debates rage around the types of skills and formation required to perform this role. It is often through consortia that curators discuss what counts as expertise in bioinformatics, and many consortia organize training workshops for aspiring curators. Last but not least, consortia promote the interests of curators and their rights as researchers, much as a workers' union would do. The status of bioinformatics within biology is on the rise and is increasingly intertwined with the emergence of "data science" as a branch of research devoted exclusively to the handling of data.[16] Among many other factors, to which I will return below, this new-found popularity is due to the visibility obtained by consortia vis-à-vis scientific institutions, journal editors, learned societies, funders, publishers, and policy makers.

This latter set of actors, including all who have institutional responsibilities toward the management and dissemination of data, constitutes the second group of interest to bio-ontology consortia. It is through effective communication with these actors, whom I will call *data regulators*, that consortia acquire the regulatory power needed to influence data handling practices across the life sciences. For instance, several consortia are engaged in dialogue with industry in an attempt to align data classification practices in that context with the practices characterizing publicly sponsored research. In addition, the GO Consortium has started collaborations with the editors and publishers of top scientific journals, which now require their authors to make use of GO labels when submitting a paper for peer review.[17] This mechanism forces experimenters to engage with GO and use it for data dissemination, which curators hope will enhance their understanding of (and interest in) bio-ontologies and thus their ability to provide feedback to database curators. Furthermore, consortia provide a platform for curators to discuss how their views can be voiced to politicians and funders in ways that are coherent and effective. The establishment of the National Institute for Biomedical Ontology, for instance, was a direct result of the lobbying of members of OBO and constituted a major step toward greater recognition and financial support of curation activities in biology by the US government and European funding bodies.[18]

The third group whose ability to intervene on data dissemination is massively increased by consortia is, of course, the vast and diverse community of *data users*. Membership of consortia attempts to transcend national culture, geographic position, and disciplinary training, thus providing a space for discussion regardless of affiliation or location. This puts consortia in a good position to foster communication between curators and users, at least in situations where enough resources are available for scientists to engage in such discussion as equal partners.[19] An effective mechanism for this is the so-called content meeting, a workshop set up by curators to discuss specific bio-ontology terms in the presence of experts from several related fields. For instance, the GO Consortium organized a content meeting at the Carnegie Institution's Plant Biology Department in 2004, in which the GO terms "metabolism" and "pathogenesis" were critically discussed and redefined through discussions among curators and experts in immunology, molecular biology, cell biology, and ecology. Similar to this are "curator interest groups," in which users are invited to provide feedback on specific ontology contents and online discussion groups coordinated through wikis or blogs. Some consortia have also discussed implementing peer review procedures on each process of data annotation, for instance by asking two referees from the bench to assess the validity and usefulness of specific bits of curators' work. This procedure, though time consuming, might become popular, especially for complex annotations relating to pathways or metabolic processes. Last but not least, consortia forcefully promote user training, through both workshops at conferences and in home institutions and by pushing the insertion of bioinformatic courses within biology degrees, often already at the undergraduate level. More than any other factors, this influence on science education is likely to reduce the gap in skills and interests currently separating curators from users.

Bio-ontology consortia have emerged from the deliberate, reflexive efforts by curators to collaborate with users and scientific institutions toward the improvement of data dissemination processes. By facilitating interactions among data curators, users, and regulators, consortia promote the use of bio-ontologies as efficient use of data sharing. This in turn increases their ability to enforce collaboration around data practices and to contribute to the broader shift in science policy toward the use of data infrastructures as main tools for collaborative and interdisciplinary work.[20] This brief analysis of the emergence and social role of bio-ontology consortia demonstrates how the process of labeling data for travel is both an outcome of—and a platform for—the regulation of data dissemination. Databases can only help to increase the fluidity of scientific communication and the circulation of related resources within an adequate institutional setting, and consortia play a significant role in developing, maintaining, and legitimizing practices of data dissemination.

2.2 Centralization, Dissent, and Epistemic Diversity

I have shown how consortia operate as collectives, gathering actors involved in the performance of research and encouraging them—sometimes forcing them—to interact with each other. Regulatory measures around data sharing procedures thus emerge from consensus achieved through frequent confrontation between different parties. As pointed out by Alberto Cambrosio and collaborators, this kind of consensus does not have to concern all aspects of scientific work but rather the modalities of use of technologies that need to be shared across large and diverse communities.[21] Furthermore, it is conceived pragmatically as a temporary achievement, which needs to be frequently challenged and revised through the expression of diverse viewpoints. Indeed, the explicit formulation and discussion of dissent among epistemic cultures is often recognized as necessary to secure the efficacy and relevance of conventions such as data classifications. Like many other organizations devoted to the regulation of data infrastructures, bio-ontology consortia construe themselves as platforms to voice the epistemic diversity characterizing local research cultures. The coordinators of the OBO consortium acknowledge the diversity of expertises and stakes in research in the life sciences, as well as the need for data users to work within their own networks and local epistemic cultures, as follows: "Our long-term goal is that the data generated through biomedical research should form a single, consistent, cumulatively expanding and algorithmically tractable whole. Our efforts to realize this goal, which are still very much in the proving stage, reflect an attempt to walk the line between the flexibility that is indispensable to scientific advance and the institution of principles that is indispensable to successful coordination."[22]

By fostering consensus through the acknowledgment of epistemic pluralism, consortia are making a political move: they are proposing themselves as *regulatory centers* for data sharing processes. As I have shown, they play a central role in shaping the expertise required to build and maintain tools for data sharing. They are also centralizing procedures, as demonstrated by their attempts to establish common rules for bio-ontology development. And they promote common objectives for the scientific community, such as the willingness to integrate the tools used to share materials and resources from which knowledge can be extracted (resources such as data but also tissue samples, in the case of biobanks, or specimens, in the case of natural history collections or stock centers for model organism research). In her reflections on pre-GO attempts to integrate community databases, Lewis emphasized the importance of setting up a common focus for collaborations around data dissemination, which otherwise risk ending in failure.[23] Furthermore, unity of purpose fosters the impression that consortia are prepared to take responsibility for the postproduction management of data, thus filling a regulatory niche that few other organizations have yet attempted to fill.[24]

This type of centralization has several epistemic and institutional advantages. It enhances the power of labels and standards to cross boundaries; it enables constructive dialogue between curators and users of databases; and it favors the cooperation between academia, governmental agencies, and industry toward the disclosure and dissemination of data. At the same time, centralization processes of any kind have long being associated with the imposition of values, norms, and standards by one group over others, and thus with a reduction in epistemic diversity.[25] This is what OBO curators recognize when discussing the difficulties of "walking the line" between flexibility and stability in regulating data sharing in the quote above. Their greatest challenge, as they openly recognize, is to implement forms of scientific governance and decision making that are fueled by a diversity of inputs and promote a diversity of uses and perspectives. As long as consortia keep up their efforts to walk that line, bio-ontologies have a chance to develop as a valuable labeling system for data. The solution to the classification problem is therefore institutional as much as it is technological. Bio-ontologies provide both the means and the platform to constantly update classificatory categories, while at the same time attempting to cultivate the epistemic diversity that is needed for data to be widely shared and reused.

Emphasizing the importance of centralization may sound counterintuitive in the face of decentralized social media such as wikis, which are a hybrid of crowdsourcing initiatives that rely on Internet users for contributions and community annotation tools through which anyone can help to curate data for dissemination.[26] These initiatives are crucial avenues for the involvement of users in the development of databases, and many consortia are seeking to exploit them. However, they often end up complementing, rather than substituting, the work of organizations such as consortia. This is because the existence of some common terminological standards and centralized co-ordination are necessary requirements for their functioning.[27] This is widely acknowledged even by defenders of the role of local agency in designing tools for data dissemination. Consider for instance the following statement, extracted from a review of the usefulness of decentralization in the development of bio-ontologies: "Local agency, when incorporated into the wider design concept, is increasingly seen as a resource for maintaining the quality, currency and usability of locally generated data, and as a source of creative innovation in distributed networks."[28] This quote underscores the idea that interventions by database users are essential to the effective functioning of bio-ontologies, and yet this can only happen in the presence of a "wider design concept." This common vision is what consortia help to achieve. Just as Google thrives on the efforts and participation of its users, while at the same time providing an adjustable but nevertheless top-down framework for such interactions to happen, online databases use a common framework to capture the interventions of data producers and users.

As widely noted within the social sciences, the recognition of cultural diversity and of the need to facilitate intercultural communication are key characteristics of governance today. The life sciences are no exception. The community involved in this type of research has never been so large, so geographically dispersed, and so diverse in motivations, methods, and goals. In such a context, efficient channels of communication are important ways to "make order"[29]—that is, to establish a structure through which individuals and groups can interact beyond the boundaries imposed by their location, disciplinary interest, and source of funding. Consortia play an important role in the management and distribution of labor and accountabilities relating to data journeys, as well as in the regulation of data ownership. By making access to bioinformatic tools conditional on the adoption of specific data dissemination practices, curators use consortia toward "the co-production of technical and social orders capable of simultaneously making knowledge and governing appropriation."[30] Thus consortia serve a regulatory function that is complementary to legal frameworks, which are typically constructed by nonscientists and imposed by state agencies rather than emerging from the experiences and expertise of practitioners.

It remains to be seen whether consortia will manage to voice epistemic diversity and highlight local agency in ways that help to push biological and biomedical research in new, productive directions or whether their increasing power and size, coupled with the overwhelming amount of data and publications to be processed, will make it increasingly difficult for them to be receptive to the diverse needs of researchers. Most attempts made so far to centralize access to biological datasets have failed due to the inability to devise labels that are intelligible and useful to prospective users, as well as immediately and efficiently applicable to large masses of data. A recent example of this kind of failure is the Cancer Biomedical Informatics Grid (caBIG), created in 2003 to function as an online portal linking together datasets gathered by the research institutions and patient care centers under the purview of National Cancer Institute (NCI) in the United States. The initial goal of caBIG curators was to identify and integrate all existing data collections generated by cancer research across a wide range of institutions. Despite the vast funding and sophisticated expertise involved in realizing this goal, the task proved unmanageable, and caBIG was increasingly critiqued for its limited usability. In 2011, the NCI reviewed the achievements of caBIG and concluded that "the level of impact for most of the tools has not been commensurate with the level of investment,"[31] a stark evaluation that led first to a decrease in funding and eventually to the demise of the program in 2013. A different example is provided by TAIR, The Arabidopsis Information Resource that I presented in chapter 1 as a key model organism database for the plant science community throughout the 2000s. TAIR came under fire around 2007 because of its perceived unsustainability and difficulties in keeping up with users' demands. Its funding was curtailed in 2008, and it was revamped as a subscription service in 2013. Despite the problems, discontinuing TAIR completely was never an option, because a substantial group of plant scientists insisted on the value of the resource and the necessity of maintaining it to the highest possible standards. Thus the restructuring of TAIR could be seen as proof of its success: the resource was so valuable to biologists that it had to be kept, but the increasing demands of users, as well as the increasing volume and types of data stored therein, meant that adequately maintaining TAIR required a level of funding and technical expertise that could only be found through association with broader initiatives such as iPlant and complementary databases such as Araport.[32]

These cases exemplify some of the opportunities and dangers associated with data packaging activities. The centralization of power over what counts as good packaging gives certain groups the opportunity to shape the choice of labels according to their own preferences and interests. In the absence of mechanisms diffusing such power, such as consortia with members located across countries, institutions, and disciplines, this centralization may entail a loss of receptivity to what database users require. This is what happened with caBIG, where an excessively top-down management imposed the use of specific standards and labels without adequate consultation with—and feedback from—prospective users. At the same time, good packaging also requires adequate resources, without which it is impossible for curators to keep up with users no matter how sophisticated their organization and feedback mechanisms are. This is the situation that TAIR encountered during the second half of its existence, where the lack of financing and relevant expertise made it difficult for curators to handle increasingly complex research needs, despite their ongoing attempts to engage with users. Other examples in the history of science point to the difficulties involved in coordinating diverse interests, values, terminologies, and constraints involved in the relevant research areas and mustering enough financial and human resources to do so in the long term.[33] Another source of worry is the fact that model organism databases, like most widely used data infrastructures, are grounded in Anglo-American science and funding structures, physically located in the richest and best resourced spots on the globe, and use the English language as a de facto lingua franca for their efforts. While English has become the closest thing to an international language for science, this choice still leaves out efforts of data production and use that are carried out in other languages and hence relate more closely to epistemic cultures and scientific traditions that never found visibility in the Anglo-American world. This imbalance of power and representation in the framing of data journeys is perhaps the most worrying downside of the centralization of data packaging efforts discussed in this section. It is hardly possible to support data journeys on a global scale without relying on common infrastructures, and yet the positioning and use of these infrastructures is grounded on existing inequalities that include, but are not limited to, assumptions about the excellence and epistemic promise of specific research traditions. In the next section, I question the very idea of "global" data dissemination by considering how debates on data management within biology intersect with the increasing visibility of the Open Data movement within and beyond the realm of research—a subject that readily illustrates how data packaging activities are situated in relation to not only the rest of sciences but also society at large.

2.3 Open Data as Global Commodities

The role of data as evidence for scientific claims makes them into public objects, which, at least in principle, can and should be widely scrutinized to assess the validity of the inferences drawn from them. And yet the vast majority of scientific data generated in the second half of the twentieth century have only been accessed by small groups of experts; and very few of those data, selected in relation to the inferences made by the scientists who analyzed them, have been made publicly available through publication in scientific journals. This management of data dissemination is tied to a view of scientific knowledge production as an esoteric and technical process, where even trained researchers become so specialized as to be unable to assess data produced by fields other than their own. Within this view, scientists invest time and effort in scrutinizing data produced by colleagues only when they have reason to doubt their interpretation or suspect foul play; and concerns with data production and interpretation, including issues associated to the management of big data in the biological and biomedical sciences, remain remote from civil society.

Since the start of the new millennium, the Open Data movement has challenged this technocratic way of conceptualizing practices of data sharing and their political, social, and economic significance. The movement brings together scientists, policy makers, publishers, industry representatives, and members of civil society around the world who believe that data produced by scientific research should be made publicly accessible online and freely usable by anyone. Participants in the Open Data movement embrace the opportunity provided by digital technologies to circulate data in real time, no matter their geographic origin. They typically advocate that data can and should travel beyond the specific settings in which they are generated, because free and wide data dissemination enhances the chance that people who have not been involved in their production (whether they are scientists or not) will contribute to their interpretation. Accordingly, they agree on the vision of scientific knowledge fostered by the Royal Society and the Organisation for Economic Co-Operation and Development (OECD), according to which the centralized collection and mining of datasets gathered by research communities across the globe maximizes the chances of spotting significant data patterns, and thus of transforming data into knowledge.[34]

This methodological and conceptual shift in the scientific status of biological data has developed hand in hand with scientific institutions geared toward encouraging and facilitating data dissemination, such as bio-ontology consortia, and fits well with preexisting commitments to data sharing in at least some parts of the biological community, such as model organism biology. However, over the last decade, funding agencies such as the National Institutes of Health, National Science Foundation, European Research Council, and Research Councils UK (as well as several others, including all those affiliated with the Global Research Council) have started to endorse this perception of how data should be managed in ways that are more explicit and forceful than ever before. These institutions support the idea that the efficient reuse of data presupposes data dissemination on a global scale, thus settling on a "politics of coordination" of the type outlined by Dirk Stemerding and Stephen Hilgartner.[35] They are promoting Open Data as key to the advancement of basic research and its translation into applications with immediate social impact, such as therapeutic or agricultural innovations[36]—and are pressuring their grantees to release data to public databases. This move affects how scientists set up their research and how they measure and develop their outputs, whether or not they agree with the principles of Open Data. This is particularly notable given the high levels of resistance manifested by researchers to the idea of sharing results indiscriminately. Even aside from the important issues of privacy and surveillance arising from the dissemination of human data,[37] there are concerns about intellectual property, potential loss of competitive advantage over other groups working on similar topics, difficulties in tackling discipline-specific approaches to data production and use, and trade-offs of spending time and resources to disseminate data of sometimes dubious quality.[38] Further, many scientists point to twentieth-century science as a time of enormous achievements, which were largely obtained despite the lack of public access to data.

In the face of such uncertainties and resistance, why are funding bodies insisting on Open Data as crucial to twenty-first-century research? A standard answer, often endorsed by journalistic accounts of the power and promise of big data, points to Open Data as a way for scientists to exploit the emergence of new technologies, such as genome sequencing and Internet-based social media. However, the emergence and political impact of the Open Data movement is not a mere consequence of technological advances in data production and communication, nor are its implications restricted solely to science. In what follows, I argue that the scientific concerns underlying the Open Data movement need to be evaluated in relation to at least four other sets of factors. First, Open Data provides a common platform for scientists, scientific institutions, and funders to discuss and tackle the practical difficulties involved in allowing data to travel and be reused. Second, it feeds into concerns with transparency, legitimacy, and return on investment on the part of political and funding bodies. Third, it aligns with the challenges posed by the globalization of biomedicine to new parts of the world and the resulting infrastructural fragmentation, geographical dispersion, and diverse characteristics of research processes. And fourth, it exemplifies the embedding of scientific research in market logics and structures.

Let us start with the importance of *bringing together a variety of groups to tackle the difficulties involved in making data travel.* As I already discussed, data journeys depend on the existence of data regulators: well-coordinated networks of individuals, scientific communities, companies, and institutions that take responsibility for developing, financing, and enforcing adequate infrastructures. In response to the technical demands involved in these efforts, companies like IBM and leading universities such as Harvard, MIT, and Berkeley are pioneering training provision in data science. This is proving to be a real challenge, since establishing data science degrees requires consensus on what types of expertise these researchers should be equipped with and what role they should play in relation to existing disciplines. Should curators of biomedical databases have both biological and computer programming skills? And if so, in which proportion and to which extent? Analysts all over the world, regularly convening through working groups of organizations explicitly dedicated to networking and research around data handling strategies (such as the Research Data Alliance and CODATA), are struggling with such questions.[39] Furthermore, the heightened status of data science also requires institutional and financial backing, to guarantee that data scientists have the status and power needed to develop and implement their contributions. As evidenced by the case of bio-ontology consortia, the resources and skills required to achieve such coordination are clearly not only technical but also social. Thus scientific institutions are starting to incorporate mechanisms to recognize and reward the contributions of data scientists. Industries are also increasingly supporting bioinformaticians as key contributors to R&D, and the publishing industry is trialing new journals devoted to documenting data dissemination strategies.

These new infrastructures and social systems are in turn embedded within wider political and economic landscapes, which brings me to the second set of factors underlying the push toward Open Data: the emphasis placed by public institutions responsible for science funding, often under pressure from national and international policy, on *fostering public trust in science as a source of reliable knowledge and thus as a legitimate source of information*. Promoting the transparency and accessibility of science is particularly important in the face of its technical nature, which makes it difficult to comprehend for the vast majority of citizens, and the general disillusionment surrounding technoscientific achievement particularly in the West. Cases in point are the controversies raging over the safety of genetically modified foods in Europe and the conviction of earthquake specialists in Italy, who were accused of failing to predict the 2009 devastation of the city of L'Aquila.[40] Perhaps the most blatant case of public mistrust in science is the controversy following the public release of e-mails exchanged by researchers at the Climatic Research Unit of the University of East Anglia in 2010—an episode often referred to as Climategate.⁴¹ This was a case where a perceived lack of transparency in how climate data were handled fueled social mistrust in the scientific consensus on global warming, which in turn affected public support for the implementation of international measures against climate change. To avoid being confronted with such situations, many national governments and international organizations support the free circulation of data in the hope that it will increase the transparency and accountability of scientific research—and thus, potentially, its trustworthiness and social legitimacy. Similarly, the Royal Society has pointed to Open Data as an opportunity to prevent scientific fraud and increase public trust in science by disclosing the evidence base for scientific pronouncements.⁴² Unsurprisingly, these expectations are generating heated debate on what should be done to make Open Data intelligible to those who lack the skills and background knowledge to understand the research being carried out, particularly given the inconceivably large amounts of information that are being released as a result of Open Data policies—a stark reminder that revealing can itself be an effective way to conceal.⁴³

The increasing reliance on, and support for, online data dissemination is also intimately related to *the globalization of science beyond traditional centers of Euro-American power* (my third point). Open Data are implicated in transforming the geographies of science and its relation to local economies, as illustrated by the rise of centers of biomedical research excellence in the Global South. Centers such as the Beijing Genomics Institute, one of the powerhouses of contemporary data production in biomedicine, interact with researchers across the world largely through digital means and do not see themselves as requiring the support of extensive local or even national research infrastructure and traditions. Information technologies enable them to quickly learn from results produced elsewhere and contribute their own share of data to international databases and research projects, thus gaining visibility and competing with established programs in the United States, Japan, and Europe. Nations that have not figured as prominent producers of scientific knowledge throughout the twentieth century, such as China, India, and Singapore, are devoting increasing financial support to research that interfaces with English-speaking science in the hope of attracting a highly skilled workforce and boosting their industrial productivity and economic prospects. The conditions of possibility for data journeys, and the diverse motivations, stakes, and incentives underlying data donation to online databases, illustrate how widespread data dissemination has created new forms of inclusion and connectivity—and, by the same token, new forms of exclusion and disconnection, which put in question the very idea that data uploaded on the Internet are thus instantly and effortlessly transformed into a "global" entity that is equally accessible and usable from any location. Indeed, one might think that laboratories in poor or underfunded regions would strongly support Open Data, as it makes data produced within expensive facilities in developed countries accessible to them, thus raising their own chances of producing cutting-edge science. Conversely, it could be expected that rich laboratories, whose access to cutting-edge technologies and vast resources gives them competitive advantage, would be reluctant to donate data—particularly as donation requires additional labor. However, taking account of the considerable resources and diverse expertise needed to transform data into new knowledge helps to provide a more realistic view on the benefits and costs of data dissemination and thus of the stakes for researchers working in very different research environments. It turns out that underfunded laboratories are often uninterested in Open Data: they may struggle to access online resources, adequate bandwidth, appropriate expertise, and computers powerful enough to analyze data found online (not to speak of relevant linguistic and cultural skills); they may not have the capability to become involved in developing standards and tools for data dissemination; and they often refuse to donate their data, which constitute their most important assets in their effort to produce publications.[44] By contrast, many rich laboratories have found that investing a small part of their large resources in data donation offers the opportunity to participate in international networks and receive help with data analysis, thus further increasing their own prestige, visibility, and productivity. Even major pharmaceutical companies like GlaxoSmithKline and Syngenta have recently started to contribute substantially to the development of public databases in the hope of outsourcing their R&D efforts, improving their public image, and gaining from the availability of data produced through public funding.[45]

This brings me to the last set of factors that I wish to discuss as relevant to explaining the prominence of Open Data as a contemporary social and scientific movement. This is the extent to which data dissemination practices are embedded into globalized political economies. *The very conceptualization of scientific data as artifacts that can be traded and circulated across the globe and reused to create new forms of value is indissolubly tied to market logics, with data figuring as objects of market exchange.* National governments and industries that have invested heavily on data production— through the financing of clinical trials or genome sequencing projects—are keen to see results. This requirement to maximize returns from past investments, and the urgency typically attached to it, fuels the emphasis on data needing to travel widely and quickly to create knowledge that would positively impact human health. This is evident when considering the extraordinary expectations linked to the Human Genome Project and its potential to contribute to medical advances. Indeed, the allure of big data lies precisely in the impossibility of predicting and quantifying their potential as evidence in advance. If we were able to predict exactly how a specific dataset could be used in the future, and thus which data should or should not be widely disseminated, we would not need Open Data in the first place: the point of free and widespread data dissemination is that one never knows who might be able to view which data and see something new in them, or indeed whether such fruitful use of data is at all possible. As demonstrated by Mike Fortun's analysis of the deCODE case in Iceland, the opportunity to circulate and reinterpret data is unavoidably couched in vastly promissory terms, and it is hard to differentiate a priori between a fruitful data sharing initiative and one that is unlikely to yield scientific insight.[46] This makes financial investment in this area both risky and potentially rewarding, making it possible to inflate or deflate expectations to suit the dynamics of venture capitalism. At the same time, this sanctions the idea of data as global goods in the sense of being essentially placeless entities, which can, at least in theory, be transported, valued, and used anywhere. This in turn challenges existing notions of property, privacy, and effective communication in industry, government, and civil society. Many pharmaceutical companies welcome the opportunity to access personal information unwittingly circulated by citizens who are not aware of its value as data for medical research—a move widely disputed by legal scholars, advocacy groups, and medical associations as an infringement of privacy.[47] The dissemination of data of relevance to innovation in food security or bioenergy, such as molecular data on plants and plant pathogens, is similarly plagued by uncertainties about intellectual property, particularly in cases of public-private partnerships between governmental agencies and companies such as Monsanto or Shell. It is still unclear how fundamentally data journeys will transform industrial practices and relations concerning intellectual property, but it is notable that such an option is being actively considered.

Kaushik Sunder Rajan and Chris Kelty have shown how free data access has greatly helped to maximize exchange and downstream capital flows.[48] I have shown how data mobility—and thus their very status as global goods—is not a given: it requires human resources and capital, and when those are absent, it can generate inequalities and exclusions. Even the most successful initiatives are confronted with the rising costs involved in maintaining and expanding data infrastructures in the long term and are struggling to produce sustainable business plans for their activities.[49] Indeed, the European Commission has denounced the costs associated with funding the current plurality of online databases in biology as unsustainable in the long term and is pushing for increasing coordination of facilities and standards for data dissemination.[50] The National Science Foundation, which has funded many successful databases at the turn of the millennium (including most pertaining to model organisms), is also attempting to rationalize its investments in this area and is asking database curators to develop self-sustaining business models.[51] Thus not only does the widespread dissemination of biological data challenge established ways of producing, controlling, and using scientific knowledge; it also plays a key role in mediating global market exchange and international politics, achieving a prominent social, and economic significance well beyond science itself. A critical assessment of the significance of the Open Data movement for contemporary society needs to take account of all these factors, which foreground the indissoluble ties of scientific research to global political economy.

2.4 Valuing Data

I have argued that the availability of new technologies to both produce and disseminate data is far from sufficient to explain the current popularity of Open Data and data-centric research. The development of technologies and expertise for the care of data, not to mention their production and use to create new biological knowledge and interventions, is made possible by the availability of institutions that help define the financial value of data as commodities and the conditions under which data can be made to travel around the globe. These institutions range from bottom-up scientific initiatives, as exemplified by bio-ontology consortia, to full-blown international agencies, like the European Research Council and the OECD, to national governments and corporate industry. In closing this chapter, I want to stress how taking this complex institutional landscape into account enriches our understanding of what it means to *value* research data, which in turn enhances our understanding of the motivations and incentives fueling data journeys.

In my use of the term "value," I wish to capture the modes and intensity of the attention and care devoted by given individuals, groups, or institutions to given objects or processes, and the motivations underlying such attention and care, which can include a wide range of interrelated concerns.[52] Thus data are imbued with *scientific value* whenever they are regarded as indispensable evidence for knowledge claims, which involves the exercise of attention and care in recording, storing, and maintaining the data. The previous chapter has documented the role played by packaging procedures in making data travel so as to develop and expand their scientific value vis-à-vis future knowledge production activities. In this chapter, I have shown that to make it at all feasible for data to travel and increase their scientific value, market structures and political institutions need to acknowledge their value as political, financial, and social objects too. In other words, scientific data can and often do have value well beyond the evidential. They can have *political* value, for instance as tools to legitimize or oppose governmental policies, or as trade currency among national governments, lobby groups, social movements, and industries. They can have *financial* value, since the increased mobility of data is unavoidably tied to their commodification and the extent to which they are expected to generate economic surplus, as I discussed above. And they can have *affective* value. Despite the increasing pressure to go public, data are often perceived as "personal" or "private"— either in the sense that they are extracted from organic materials or behaviors belonging to a specific person or group, thus somehow capturing part of its identity, or in the sense that they are produced through the long-term efforts of a team of researchers, who therefore feel ownership over them and to some extent identifies with them as a creative, original achievement. The affective value of data is particularly prone to be underestimated by public discourse on open and big data, and yet it plays an important role in explaining why the most easily disseminated data types in biology are produced through high-throughput technologies such as sequencing. Once the appropriate machine is available, these data are the easiest to produce and thus also to give away, as there is little affective investment in them. They are very different from cases such as Franklin's crystallographic images of DNA, the production of which required abundant ingenuity, resources, and time investment and whose unauthorized reuse by Watson and Crick gave rise to one of the most infamous authorship disputes in twentieth-century biology.

This multifaceted identity of data has complex, and sometimes paradoxical, implications. One is the potential friction between affective and scientific ways of valuing data. Many champions of the Open Data movement, who wish to free data from ownership claims and property regimes, view the affective value of data as a threat to their scientific value. In their view, for data to circulate widely, and thus generate the kind of data journeys that maximizes their value as prospective evidence, data owners need to free their data from any conditions that may limit their ability to travel. And yet there may be good reasons to retain an affective attachment to data and keep at least some data away from the public eye. Blatant cases are ones concerning the infringement of patient privacy, where data being disseminated may be used to cause harm to individuals or groups,[53] or those where data access has security implications (such as the much discussed cases of data on genetically engineered diseases, which may be used to develop bioweapons[54]). Perhaps more significantly, many researchers feel that individuals who have not participated in the production of certain kinds of data are not in the position to evaluate their quality and significance as evidence and that making such data travel could harm scientific progress by encouraging misleading interpretations. Within this vision, the affective and scientific value of data are tied together, rather than being at odds with each other. For instance, many researchers investigating gene expression argue that metabolomic data, a key source of evidence in the field, are extremely sensitive to the environmental conditions in which they are originally produced and it is therefore pointless to try to integrate metabolomic data acquired from experiments carried out around the world on different materials and under different conditions.[55]

Another potential source of friction concerns the financial and political value of data. I argued that data journeys fuel some key tenets of capitalistic production, with its emphasis on excess, growth, and the role of technology in delivering social goods—a worldview shared by industry and governments alike, particularly in the developed world. And yet, at the same time, the Open Data movement attempts to demarcate science as a normative space where some resistance to the appropriation of the market can and should happen and where principles such as transparency and accountability can trump vested interests and biased interpretations—thus supporting the political use of data-centric science as a harbinger of truth in the context of highly charged governmental decision making.[56] In response to this, several corporations are developing strategies geared toward maximizing both their financial gain and their reputation as transparent and accountable scientific organizations. GlaxoSmithKline, for instance, has maintained an online database called Clinical Study Register since 2004, which is used to disclose results from their clinical trials to honor "the importance of publicly disclosing this research"[57]—a clear nod to the political value of data for legitimizing claims, actions, and in this case, products. Notably, this database is not a complete compendium of all results ever obtained in GlaxoSmithKline clinical trials. Rather, it contains selected datasets and summaries[58]—a policy that safeguards the quality and intelligibility of the information that is released while also providing the company with the opportunity to be selective about which data are disseminated.

I hope to have shown that no matter what kinds of data, infrastructures, and communities are involved, every time that scientific data embark on a journey, different ways of valuing data are bound to come into play.[59] How such standpoints, and the potential tensions between them, are negotiated shapes the travel of scientific data. Recognizing their existence is therefore crucial to understanding how data-centric science works and why it is acquiring such prominence at this particular point in time. Importantly, this recognition does not come at the expense of acknowledging the scientific, technical value of data. Assessing how data journeys can contribute to the development of knowledge is a difficult exercise—one that crucially involves scientific expertise in assessing the quality, reliability, and potential significance of data. Try searching a publicly available biological database without possessing relevant biological training: the results yielded by the search engine will most likely look meaningless. Scientific data are quintessentially technical artifacts, whose interpretation depends on the extent to which the people who use them are able to assess the conditions under which the data were produced in the first place—a point that grounds my philosophical account of the nature of data as research components, which I detail in the next chapter.

What is perhaps less obvious is that the technical nature of scientific data sits very well with their value as political, financial, cultural, and social entities. What has propelled data into becoming protagonists of contemporary biomedicine is precisely their complex status as at once local and global, free commodities and strategic investments, common goods and grounds for competition, potential evidence and meaningless information. The vision underlying the Open Data movement is that data risk remaining meaningless if they are prevented from traveling far and wide, and that travel endows data with multiple forms of value. For data journeys to extend across a variety of contexts, ranging from academic labs to industrial development departments to policy discussions, data need to be viewed as interesting by all actors involved. Given the diversity of stakeholders in the production, dissemination, and reinterpretation of biological data, it is no wonder that motivations and incentives behind all these efforts should diverge widely. Broadening our analytic gaze in this way helps to bring home the argument I defended in chapter 1—namely, that flexibility to multiple uses and future scenarios, as well as to the diverging interests of potential users, is crucial to the success of databases in enabling data journeys and thus to the future of knowledge production in biology. Unravelling the characteristics of data-centric biology is at once highly significant in view of current social, political, and economic concerns and also strongly dependent on a balanced assessment of constraints and decisions internal to scientific reasoning and the broader landscape of opportunities, demands, and limitations within which researchers operate. A critical assessment of the epistemological significance of data-intensive science needs to take account of all these factors, which foreground the indissoluble ties of biological research to national and international political economies.

Part Two: Data-Centric Science

3

What Counts as Data?

The debates on big data, data-intensive research, and data infrastructures documented in the previous chapters have re-ignited social and cultural interest in what counts as data and under which conditions data are transformed into knowledge. In this chapter, I propose a philosophical perspective on scientific data that makes sense of these developments. I conceptualize data as tools for communication, whose main function is to enable intellectual and material exchanges across individuals, collectives, cultures, nations, and—in the case of biology—species, and whose mobility across these groups is a hard-won scientific achievement.[1] This constitutes a novel perspective within the philosophy of science, whose students have so far emphasized the man-made, situated nature of data production and interpretation but have paid little attention to the ways in which data are shared after they are first generated and thus to the challenges and ingenuity involved in strategies to make data travel. I will also insist that data are material artifacts whose concrete characteristics, including their format and the medium through which they are conveyed, are as relevant to understanding their epistemic role as their social and conceptual functions. This position reflects a broader perspective on scientific epistemology that emphasizes its processual and embodied nature and seeks to understand science by studying the practices and instruments through which research is carried out.[2]

Within this view, unraveling the conditions under which data are disseminated is crucial to understanding what counts as knowledge and for whom and to assessing the epistemic value of various outputs of research, whether they be claims, data, models, theories, instruments, software, communities, and/or institutions.

The chapter starts with a brief review of debates around data within the philosophy of science. I then introduce my characterization of data as a relational category applicable to research outputs that are taken to provide evidence for knowledge claims. I argue that data do not have a fixed scientific value in and of themselves, nor can they be seen as mind-independent representations of given phenomena. Rather, they are defined by the *evidential value* ascribed to them at specific moments of inquiry—that is, the range of claims for which data can be considered as evidence. This position makes sense of the crucial role played by data packaging strategies within data-centric science, where data are perceived as potentially able to support claims that scientists have yet to formulate, and indeed may never formulate—in short, as having a potentially inexhaustible evidential value. Within data-centric research, data are not disseminated for the sake of preserving them as intrinsically valuable objects, as would be the case in any collection exercise, but rather for the sake of their repurposing to serve new scientific goals. This is what packaging strategies seek to achieve by creating conditions under which a given set of objects can function as data, such as making those objects widely accessible and making it possible to retrieve, visualize, and manipulate them at will. Packaging strategies play a crucial role in expanding the evidential value of data—and in so doing, they affect whether or not a given set of objects counts as data in the first place and for whom.

This view of data may seem counterintuitive from an ontological viewpoint. Processes of packaging can and often do change the format, medium, and content of data as they travel across databases and laboratories and are repurposed for a variety of evidential uses. Given this mutability and the functional nature of this conceptualization, how can data retain integrity as the "same" objects throughout their journeys? Under which circumstances does a given set of objects stop being data and become something else—materials, models, facts, noise, or waste? To which extent is data interpretation tied to the specific circumstances under which they are originally produced? And how can data be distinguished from other research components, particularly models, that play such an important part in packaging processes? I address these questions in the second part of this chapter.

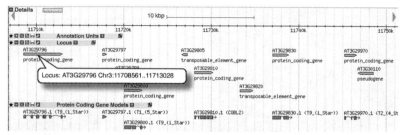

FIGURE 2 A 45kbp-wide region (kilo base pairs) on the third chromosome of *Arabidopsis thaliana*, including indications of known genes and the proteins they code, as stored in the TAIR database and viewed through GBrowse (October 22, 2015). Many more data can be optionally shown by the browser, such as community/alternative annotations, noncoding RNAs and pseudogenes, polymorphisms and transposons, methylation and phosphorylation patterns, orthologs, and sequence similarities.

3.1 Data in the Philosophy of Science

Let us start by considering some examples of what biologists regard and use as data. These may come in various shapes and formats, including the measured positions of gene markers on a chromosome (figure 2), the scattered colors indicating gene expression levels in a microarray cluster (figure 3), or the photographs taken to document different stages of embryological development (figure 4). Etymologically, the term *data* is Latin plural for the expression "what is given." These biological examples show the resonance that the idea of data as "given" has in the life sciences. These are images that can be easily construed as a starting point for scientific reasoning about a variety of phenomena, including genome architecture, the patterns of expression of specific genes, and their impact on early development. They are generated through processes of measurement and manipulation of organic samples undertaken under controlled conditions, and as such, they are taken to document, as accurately as possible given the instruments used, features and attributes of a natural entity—in these cases, an organism like a plant (figure 2) or a fruit fly (figure 4). Interpreting the scientific meaning of these figures is left to researchers, who decide whether to regard them as evidence for specific claims about phenomena on the basis of their interests, background knowledge, and familiarity with the procedures through which the data were obtained.

The importance of human agency in attributing meaning to these images points to a tension that characterizes the role of data in research, and which provides a starting point for philosophical analysis. It consists of the observation that despite their scientific value as "given," data are clearly made. They are the results of complex processes of interaction between researchers and the world, which happen with the help of interfaces such

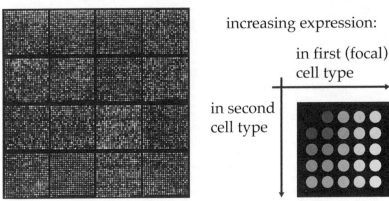

FIGURE 3 A mouse cDNA microarray containing approximately 8,700 gene sequences and an interpretation chart for the relative expression levels. Here two different mouse tissues are compared; conventionally, in cancer studies, the red fluorophore marks the cancerous cells and the green marks the "normal" cells. Researchers evaluate gene expression levels by comparing each colored dot in the microarray to a chart: if the color falls in the bottom-left triangle, expression is stronger in the second versus the first cell type, and vice versa for the top-right triangle; dots on the separating diagonal represent unchanged expression levels. (Image courtesy of the National Institutes of Health; chart produced by Michel Durinx, October 2015.)

as observational techniques; registration and measurement devices; and the rescaling, modification, and standardization of objects of inquiry (such as organic samples or even whole organisms) to make them amenable to investigation. This is the case for data resulting from experimental processes of manipulation, which in turn involve recourse to complex apparatus and procedures that embody specific interpretations of the world—an example of what Ronald Giere calls the perspectival nature of observation.[3] It is also the case for data generated outside the controlled environment of the laboratory. For example, handwritten notes or photographs of the movements of a herd of bison, taken by ethologists to document their observations of bison behavior, are conditioned by the employment of specific techniques and instruments (a certain type of camera, a small or large notebook, and specific type of pen), as well as the interests and position of the observer, who can only watch the scene from a specific angle and a given distance.[4] In the words of Norwood R. Hanson, the very act of seeing is a "theory-laden undertaking."[5]

The tension between viewing data as instances of the world and emphasizing their man-made nature has acted as a thread for philosophical discussions of scientific methods at least since the scientific revolution. For the most part, philosophers have focused their efforts toward debunking the myth of data as given rather than made. Over again, they noted that humans are too conditioned by their own assumptions and interests to be able to

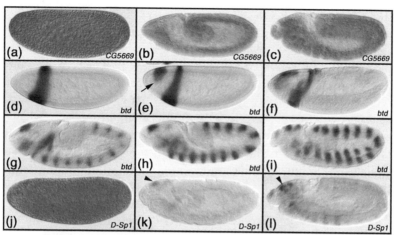

FIGURE 4 Photographs of fruit fly embryos expressing three Sp genes across different stages of development. All embryos are arranged with the anterior to the left, and arrows in the last two images show expression in the developing brain. (Nina D. Schaeper, Nikola-Michael Prpic, and Ernst A Wimmer, "A Clustered Set of Three Sp-Family Genes Is Ancestral in the Metazoa," *BMC Evolutionary Biology* 10(1) (2010): 1–18. doi:10.1186/1471-2148-10-88.)

observe the world objectively and that scientists are no exception. Scientific methods have been hailed as efficient means to moderate, and where possible annihilate, such subjectivity—an achievement that presupposes the recognition that what one takes to be a fact about the world may well be a fallacious impression generated by the senses. Accordingly, most prominent philosophers in the Western tradition have regarded the method of induction with suspicion. Knowledge may well be grounded on observations, but it is crucial to critically consider how those observations were generated in the first place, if science is to challenge (rather than simply confirm) the many mistaken assumptions that humans tend to make about the world they live in. A particularly vocal advocate of this view was Pierre Duhem, whose early twentieth-century writings inspired authors such as Hanson and Thomas Kuhn to emphasize the inevitable influence of theoretical presuppositions on the collection, selection, and interpretation of data.[6]

Debates on theory-ladenness have repeatedly clashed against the intuition that data constitute "raw" products of research, which are as close as it gets to unmediated knowledge of reality.[7] Under the influence of logical positivism and Karl Popper's falsificationism, many philosophers of science have dealt with this clash by implicitly or explicitly subscribing to two interrelated assumptions about the role of data in scientific research. The first assumption is that data are representations of the world, in the sense that they capture some features of reality so as to make them amenable to scientific study. This implies that the information content of data is fixed regardless of how they are used by researchers and defines their identity as primary empirical grounding for knowledge claims. Data may be man-made, but their features are determined by the material circumstances of their production, which makes them largely mind-independent. Indeed, what exactly data should be taken to document may not be known and needs to be uncovered through processes of analysis and interpretation, which may generate surprises and counter preformed expectations. This brings me to the second assumption: the idea that data are primarily means to test and validate theories, and scientific methods can be safely assumed to guarantee their reliability in this role. Philosophers such as Hans Reichenbach and Carl Hempel put enormous trust in the idea that controlled conditions for data collection (as in experiments and randomized trials) and sophisticated statistical analyses make it possible to evaluate objectively what information a given dataset contains and whether this is being interpreted correctly or incorrectly. How can data, understood as an intrinsically local, situated, idiosyncratic, theory-laden product of specific research conditions, serve as objective confirmation for universal truths about nature? Hempel's answer to this question is to insist that scientific methods provide "directly observable" and "intersubjectively ascertainable" observations that can be taken as irrefutable, objective facts about physical objects—and used to validate a given theory or explanation.[8]

These two assumptions form the core of what I call the *representational view of data*. This view is tied to a theory-centric understanding of scientific knowledge production, according to which the main outputs of science, and hence those most deserving of philosophical scrutiny, are those least dependent on fallible human perception; the best theories are ones expressed via universalizing mathematical axioms; and explanations are conceived as deductions, adequately adapted to the question at hand, from nomological systems. Within this tradition, whether and how scientific methods yield reliable data is regarded as an empirical matter of no consequence to philosophical analysis, which should instead focus on reconstructing the logic through which axioms and theorems are constructed and related to observations and measurements.[9] Hans Reichenbach enshrined this attitude in his notorious separation between the messy conditions in which data are produced and used (the "context of discovery"), whose study is the job of historians of science, and the neatness and rationality of the resulting claims (the "context of justification"), which constitute appropriate objects of study for philosophers of science.[10]

This theory-centric view of scientific knowledge has been challenged over the past three decades by what is sometimes dubbed the *practice turn* within Anglo-American philosophy of science. Starting from the late 1970s, an increasing number of philosophers started to pay more attention to authors such as Francis Bacon, William Whewell, and John Stuart Mill, who emphasized the fruitfulness of examining the actual features of processes of discovery, rather than their post facto reconstruction.[11] This new interest in the practical ways in which scientific knowledge is achieved was channeled in the study of scientific models, whose epistemic role has been found to vary widely depending on their concrete features and the interests and values of their users.[12] This drew attention to the exceptional diversity characterizing scientific research, which might be better investigated starting from the idiosyncrasies of specific cases rather than from an emphasis on common strategies and overarching theories.[13] It is from this practice-inspired literature that some of the most interesting contemporary challenges to the representational view of data have emerged. Within the philosophy of experiment, Ian Hacking contributed a broad definition of data as *marks* produced by human interactions with research instruments. This account portrays data as things that can be stored and retrieved and are thus made durable. By focusing on the material circumstances in which data are generated, Hacking remains agnostic about the epistemic role that data may play in scientific inquiry and emphasizes instead the constraints and opportunities provided by the manifold of formats and shapes in which data are produced in the laboratory—comprising, in his words, "uninterpreted inscriptions, graphs recording variation over time, photographs, tables, displays."[14]

Another philosopher who takes seriously the concrete features of data is Hans-Jörg Rheinberger, who grounds his views on the epistemic role of data on a detailed historical examination of experimental procedures in biology. His conclusions differ from Hacking's insofar as he does not view the marks produced by scientific instruments—which he calls "traces" or "signals"—as data. Instead, he conceives of data as the result of further manipulations of marks resulting from observation or experiments, which are performed with the purpose of storing them and making them available to others (what I have called packaging). As an example, Rheinberger points to the first DNA sequence gel produced by Fred Sanger and collaborators in 1977. The gel helps to visualize the molecular structure of the DNA sequence of bacteriophage PhiX174 by generating discrete stripes of varying lengths on a photosensitive plate. Rheinberger interprets those stripes as *traces* generated by this laboratory technique and the abstraction of these stripes into a chain of symbols standing for the four nucleic acid bases, GATC, as an example of the transformation of traces to *data*.[15] This account benefits from the success of the use of letters as symbols for nucleobases, a format that made this type of data easy to circulate and thus facilitated the rise of molecular approaches, particularly genomics, within biology. Rheinberger also explicitly builds on Bruno Latour's work on chains of inference, which highlights how the establishment of knowledge claims is grounded in the production and movement of "immutable mobiles"—that is, objects that can serve as anchors for knowledge claims thanks to their stability across contexts.[16]

Both Latour and Rheinberger recognize that the marks produced in the course of research need to be processed in order to travel, and traveling is crucial to their functioning as evidence. Taken jointly, their accounts contribute a key insight to my analysis—that is, the epistemic importance of the mobility of data and the labor required to realize it. What I do not share with them is the emphasis on stability—or, in Latourian terms, immutability. When traveling from their original context of production to a database, and from there to a new context of inquiry, biological data are anything but stable objects. The procedures involved in packaging data for travel involve various stages of manipulation, which may happen at different times and may well change the format, medium, and shape of data. This is the case even once data are presented in apparently straightforward formats, such as the sequence data used in Rheinberger's example. While the use of letters to indicate nucleobases is among the most recognizable and universally intelligible symbols in contemporary biology, biologists and curators have a vast choice of file formats in which those data could be stored and visualized— for instance, the Staden format simply provides the letters in succession, as in "GGTACGTAGTAGCTGCTACGT," while both the European Molecular Biology Laboratory (EMBL) database Ensembl and the sequence repository GenBank provide a variety of possible formats depending on the methods used to produce sequences and the users' interests.[17] As documented within the scientific literature on data curation, the subsequent travel of sequence data is certainly affected by these choices.[18] Does this process affect the ways in which data are used as evidence? I agree with Rheinberger and Latour that it does. Does it warrant a clear-cut distinction between the traces obtained through scientific intervention and their "packaged" versions, ready for travel? I think not.

Packaging happens at several stages of data travel and is often implemented already at the point of data production. For instance, as Rheinberger recognizes, increasing amounts of biological data are generated digitally, so that it is easier to handle them computationally and to submit them to online databases; and, given a choice, researchers tend to prefer instruments that can produce digital outputs over instruments that yield results in analog formats. Given the iterativity characterizing the processes through which data are produced and disseminated, trying to differentiate between the marks produced by instruments and those obtained through further manipulation seems arbitrary.[19] Scientists engage in data generation in full awareness that the outputs of that activity need to travel beyond the boundaries of their own investigation. This awareness is built into the choice of instruments used; the recording of procedures and protocols carried out in lab books; and the decisions about how outputs may be stored, shared with peers, fed into models, and integrated with other data sources. The need to package data for travel is recognized by scientists as an essential requirement for knowledge production, which underlies the planning of research, the generation of data, and their further elaboration. This undermines the idea that we can neatly distinguish between data as traces directly derived from investigation and their further translation into a variety of formats.

3.2 A Relational Framework

A better way to build on Rheinberger's analysis is to give up altogether on a definition of data that is based on the degree to which such objects are manipulated and focus instead on the relation between researchers' perceptions of what counts as data and the type and stage of inquiry in which such perceptions emerge. Data journeys do not merely affect the interpretation of data: they determine what counts as data, and for whom, in the first place. As data travel, their producers, curators, and users keep making decisions about what constitutes data in relation to the changing circumstances and aims of each stage of data journeys. These decisions are not necessarily consistent with each other and are often taken independently from each other on the basis of interests of the specific individuals involved, the materials and formats of the objects in question, the ethos of the relevant communities, existing standards for what counts as reliable data, conditions for data access and use, and shifting understandings of data ownership and value. At the same time, these decisions are typically taken with one common goal, which is to maximize the chance that data may serve as evidence for one or more knowledge claims in the future.

Building on this insight, I propose to define data as any product of research activities, ranging from artifacts such as photographs to symbols such as letters or numbers, that is collected, stored, and disseminated *in order to be used as evidence for knowledge claims*. As I highlighted in the previous chapter, data may be valued in other ways—including financial, affective, and cultural,—and this may affect their journeys. However, it is the value assigned to data as *prospective evidence* that distinguishes them from other research components. As data travels, it may not always be clear how they could be used, and it is often the case that data stored in databases is not retrieved again and is thus not employed to create knowledge. This is not a problem for database curators, as long as there is an expectation that the data may serve as evidence at some point in the future, and data are therefore handled in a way that makes them available to further analysis.

Hence data are objects that (1) are treated as potential evidence for one or more claims about phenomena and (2) are formatted and handled in ways that enable its circulation among individuals or groups for the purpose of analysis. In the case of scientific data, these groups will most likely include at least some scientists, although this is not a necessary requirement in my framework.[20] This definition frames the notion of data as a *relational category*, which can be attributed to any objects as long as they fulfill the two requirements above. What counts as data depends on who uses them, how, and for which purposes. Within this view, the specific format of the objects in question does not matter—a position that reflects the enormous variety of objects used as data by biologists, which ranges from experimental measurements to field observations, the results of simulations and mathematical modeling, and even specimens and samples of organic material. Also, there is no intrinsically privileged type of data, as judgments on which objects best work as evidence depend on the preferences of the researchers in question, the nature of the claims under considerations, the materials (such as the organisms) with which they work, and the availability of other sources of evidence. For instance, a group of behavioral psychologists may take a specific group of organisms—like a genetically engineered mice colony—to constitute data for claims such as "mice with X genetic makeup tends to exhibit behavior Y." The specimens themselves would not suffice as a source of evidence and would be flanked by other types of data such as photographs and videos of the mice, samples of their blood, their genome sequence, and observation notes made by researchers to describe their behavior. A research group with different interests and expertise, working on a different type of organism, will likely use a different combination of objects as sources of evidence. For example, a team of evolutionary biologists investigating claims such as "bacterial populations exhibit evolutionary novelties as a result of multiple mutations in their genomes" will be interested in bacterial colonies at various evolutionary stages, fitness data for several generations of these organisms, genome sequences, and photographs of the morphology of ancestral and evolved strains.[21]

A key implication of this approach is that the same objects may or may not be functioning as data, depending on which role they are made to play in scientific inquiry and for how long. This is particularly significant given the contradictions and uncertainties, evidenced in much scientific and policy literature, about how data should be defined and whether their identity changes whenever they shift format, medium, or context. Many participants in these discussions think that, despite the multiple types and uses of data across the sciences, debates concerning data-intensive science should be grounded on a context-independent definition of what data are. This arches back to a representational view of data as entities that depict a specific part of reality independently of the circumstances under which they are considered. Under this interpretation, analyzing data involves uncovering which aspects of reality they document, and their epistemic significance stems from their ability to represent such aspects of reality irrespectively of the interests and situations of the people handling them.[22] This view is incompatible with the idea that the same set of data can act as evidence for a variety of knowledge claims, depending on how they are interpreted—a feature that I take to be central to understanding the epistemic power of data as research components.[23] It also does not account for situations when objects expected to function as data cease to be treated as such, which are very frequent in research. For instance, it is common for data to be discarded as useless "noise," to be assembled for purposes other than use as evidence (e.g., museum displays, private collections), and to be stored in ways that make them impossible to retrieve for future use (e.g., inaccessible archives or databases that do not contain effective data retrieval mechanisms and are therefore critiqued by many practitioners as "data dumps" where "data go to die").

To account for these situations, I advocate defining data in terms of their function within specific processes of inquiry accounts, rather than in terms of intrinsic properties. Within this framework, it is meaningless to ask what objects count as data in the abstract. This question can only be answered with reference to concrete research situations, in which investigators make decisions about which research outputs could be used as evidence and which are instead useless in that regard. This position is purposefully not intended to help evaluate the motivations that may push scientists to consider specific objects as data, since I think that assessing these choices is a matter of scientific, rather than philosophical, competence. Rather, I am interested in how data are routinely handled within biology as a gateway to understand the circumstances under which certain objects come to be cast as evidence toward given knowledge claims. Database curators have often remarked to me that the best databases make it possible for users to play around with data by organizing them and visualizing them in a variety of ways, each of which may fit a different constellation of hypotheses, assumptions, and intuitions about phenomena. I view this as a significant insight in the epistemic role of data, which highlights the importance of their untapped potential in relation to existing and future research. What makes data into power-

"This is it, Smythe -- the Tomb
of the Lost Unstructured Data!"

Brian Moore   brianmooredraws.com

FIGURE 5 Cartoon depicting data stored in ways that make them irretrievable ("unstructured") as a "data tomb." Such satirical depictions are common within social media devoted to scientific communication and debate, particularly relating to Open Data discussion. (Courtesy of Information Week, "Cartoon: Where Data Goes to Die," December 19, 2014, http://www.informationweek.com/it-life/cartoon-where-data-goes-to-die/a/d-id/1318301?image_number=1.)

ful epistemic tools is not necessarily their provenance, format, or content, though these factors may strongly affect the scientific significance attributed to them, but rather their ability to fit a variety of lines of inquiry, which in turn depends on the ways in which producers, curators, and users mobilize and arrange them vis-à-vis specific questions and ways of reasoning.[24]

Accordingly, one characteristic that I see as essentially tied to my definition of a datum is its *portability*. Regardless of whether it is realized on the scale recommended by the Open Data movement, portability is a crucial precondition for using data as evidence. This is because the establishment of scientific claims, including the formulation of judgments about the reliability and significance of data, is widely recognized as a social activity that needs to involve more than one individual. This point resonates with historical studies of the crucial role of witnessing and communication formats in scientific research, as exemplified by Steven Shapin and Simon Shaffer's analysis of Robert Boyle's strategies to secure scientific consensus.[25] No intellectual achievement, no matter how revolutionary and well justified, can be sanctioned as a contribution to scientific literature unless the individual concerned can express her ideas in a way that is intelligible to a wider community of scientists and can produce evidence that has the potential to be exhibited to others as corroborating her claims. I therefore do not consider a situation where two or more individuals are able to observe the same phenomenon as an instance of data production.[26] Individual testimony plays an important role in science as a platform for discussion and reasoning over the potential biological significance of observations and the ways in which research should be furthered. However, having witnessed a phenomenon in the lab or in the field is not an acceptable form of evidence when trying to establish the validity of a claim as a contribution to scientific knowledge, for instance when submitting a paper to a journal. In order for any individual or group experience to be considered as evidence, objects (photographs, videos, measurements, handwritten notes) must be produced that document that experience and make it portable to other contexts. These objects are what I call data. Making data travel is therefore a necessary, though not sufficient, condition for their prospective use as evidence. If data are not portable, it is not possible to pass them around a group of individuals who can review their significance and bear witness to their scientific value.[27]

The crucial role of portability is also what leads me to characterize data as *material* artifacts, independently of whether they are circulated in a digital form or not.[28] As also emphasized by Hacking, whether we are dealing with symbols, numbers, photographs, or specimens, all data types need a physical medium in which they can be disseminated.[29] This mundane observation has a significant philosophical implication, which is that the physical characteristics of the medium affect the ways in which data can be disseminated, and thus their usability as evidence. In other words, when data change medium, their scientific significance may also shift. This is notable given the diversity of media that data are likely to encounter in their journeys. Genome sequence data, for instance, can be disseminated in a wide variety of formats and vehicles, ranging from archives to databases, journal publications, stock centers, and biobanks; and I already stressed that the format given to data when they are processed for dissemination tends to drive the types of analysis that are then carried out—and thus the type of results obtained. This observation underscores the man-made quality of data, and thus the difficulties in viewing them as objective sources of evidence. Rather than viewing this as a problem for scientific epistemology, I welcome this recognition of how deeply the characteristics of data are intertwined with the specific stages of research in which they are used. This is a key insight gained by studying data dissemination processes, particularly the packaging practices performed by database curators. Recognizing the ingenuity involved in those processes teaches us that shifts in the format and media used to disseminate data are not an obstacle to the use of data as evidence but rather an essential component of their journeys, without which they would not travel as efficiently (or at all). Highlighting the epistemic significance of the physical transformations that data undergo in order to travel is also a key motivation underlying Rheinberger's differentiation between traces and data: in this sense, my framework is compatible with his historical observations, even if not with his view of how data should be conceptualized.

The relational approach to data is susceptible to two major ontological objections. The first concerns the degree of integrity of data as physical objects. If data change format as they travel, sometimes with significant implications for how they can be analyzed and thus for their prospective evidential value, how can we account for them being in some respect "the same objects"? In other words, how can data be mutable and yet conserve their integrity, and how do we as analysts know that we are looking at the same data when they change all the time? In answering this question, I take inspiration from philosophical discussions about another entity that is widely acknowledged to be identifiable in space and time and yet changing throughout its life history: the biological individual. Organisms are clear instances of inherently unstable entities in constant transformation. Their components change all the time, as new materials are incorporated from the environment and cells and tissues die and are replaced by new ones. Their appearance, dimensions, and relationship to other entities also change continuously—not to speak of the shifts that occur if we consider a whole lineage of organisms, descending from the same ancestor, as an individual. One philosophical strategy to make sense of the ontological status of these entities is to adopt a conception of identity that takes account of the physical changes that an individual may undergo through time as it develops and transforms in relation to the environment.[30] A case in point is the notion of genidentity, introduced by Kurt Lewin in 1922 and recently adapted to the case of biological individuals by Alexandre Guay and Thomas Pradeau. Within this view, the identity of an entity X is "nothing more than the continuous succession of the states through which X goes. . . . For example, a 'chair' is to be understood in a purely historical way, as a connection of spatiotemporal states from its making to its destruction. . . . What we single out as an 'individual' is always the by-product of the activity that is being followed, not its prior foundation (not a presumed 'thing that would give its unity to this activity')."[31] This conceptualization usefully accounts for the integrity of data in time and space while also acknowledging that the physical features of data may change throughout their life history. Data are manifested as concrete entities at any one place and time and yet are continuously transformed as they travel across research sites and media. Their material manifestations make it possible to track data as they are moved around and reconstruct their overall journeys.[32] At the same time, data are not inherently stable objects, and whether or not they maintain their physical characteristics depends on how their travel is managed and on the shifting conditions under which they are viewed as evidence. The functional nature of data is perfectly compatible with this view: it provides a way to determine when data come into being and cease to be, while remaining neutral about the specific features that the objects used as data retain and/or acquire through time. Thus, from an ontological perspective, data are both a process and a substance. What counts as data at any point of time can only be determined in relation to a specific situation, and yet data exist as a mutable lineage beyond that specificity.

A second objection to the relational approach may come from philosophers who distinguish between the identity of a dataset as unique source of information (a "type") and the concrete instantiations of such dataset whenever it is copied through a multitude of media (the "tokens" of that type).[33] Within this view, types are immaterial forms whose individual instantiations may be concrete but whose unique identity is immutable and intangible. An example would be a colored dot on a microarray slate, a letter in a sequence, or a dark blurry spot in a photograph of organic tissue, which may be viewed as remaining "the same" no matter how and how often they are reproduced in screens, textbooks, or articles. Copyright law makes a similar distinction when discussing the difference between an original work and its copies and separating the intangible forms to which authorship may be attached from the specific, tangible instantiations to which property claims may be attached irrespectively of whether one is the author or not.[34] My response is that the distinction between token and type does not help to make sense of the epistemology of data. I have already noted how data do not have a fixed information content, and their form—the media through which they are instantiated—may significantly affect the ways in which they are used as evidence. Furthermore, the more data travel, the more difficult it becomes to distinguish their authors from their owners. Consider a biologist who produced a set of microarray data in her lab and deliberates on how such data should be disseminated. If she wishes to be recognized as author of the data, she may publish the dataset in a data journal or donate it to a database that mentions her as the source. This will tie her name to the format in which those data were originally produced. This format is, however, likely to change once the data start their journeys across screens, printouts, and databases around the world. Many other individuals may invest resources and ingenuity into modifying and formatting data to fit new uses, in ways that may well prove as significant to the interpretation of the data as the efforts of their creator. In such a scenario of data dissemination and reuse, it does not seem fair to single out the original producer of data as their only author, and the distinction between types and tokens becomes meaningless.[35]

3.3 The Nonlocality of Data

I now want to probe more closely the idea that the evidential value of data is defined by the context in which they are produced. Within the philosophy of science, many scholars maintain that because the specific circumstances in which data are created affect how they may be interpreted, data can never be considered independently of that context. My account challenges this inference by refocusing philosophical attention on processes of dissemination rather than production. Data interpretation certainly does require reference to the specific circumstances in which they were created, and I will discuss this at length in the next chapter. However, this does not limit the number of ways in which data can be interpreted, because their production context does not fully determine their evidential value. No matter how tightly controlled an experiment is, or how well known its targets already are to scientists, what the resulting data may end up revealing is not entirely predictable, nor can it be captured by any single claim. This is not because it is impossible to predict at least some future use of data as evidence: this is clearly the case in many instances and the rationale for planning experiments to explore given hypotheses or questions. Rather, it is because one cannot predict *all* the possible claims that data might be used as evidence for in the future; and also because one cannot predict whether data will *actually* be used as evidence for specific claims until it happens. In Duhemian terms, the evidential value of data is underdetermined—a situation that data-centric methods propose to exploit, in order to maximize the fruitfulness of data within multiple research contexts. In what follows, I discuss the implications of this insight for one of the most prominent philosophical arguments hitherto put forward for the intrinsic locality of data: James Bogen and James Woodward's account of data processing in experimental physics.

I should start by noting that Bogen and Woodward's work constitutes a crucial reference point for my account in several respects. Their analysis shares my interest in using scientists' own concerns and actions as a starting

point for understanding the epistemological role of data. As a result, Bogen
and Woodward emphasize the scale and significance of the efforts and skills
required to generate and use data, with Woodward explicitly stressing the
public nature of data as a record of scientific activities "whose acceptability
depends upon facts that can be ascertained intersubjectively."[36] Moreover,
their work has resurrected philosophical attention to data as objects that
can be straightforwardly observed and whose main function is to serve as
evidence for knowledge claims. To achieve this, they introduced a distinction between "what theories explain (phenomena or facts about phenomena)" and "what is uncontroversially observable (data)."[37] Similarly to my
framework, the main function of data is to provide evidence for claims,
but Bogen and Woodward noted these claims did not amount to theories.[38]
Rather, data serve as evidence for claims about phenomena: unobservable
features of the world that constitute "the facts for which scientific theories
are expected to account."[39] Examples of phenomena include the melting
point of lead and the charge of the electron. Within biology, comparable
cases would be the notions of gene expression, phenotype, and metabolic
pathway. Claims about phenomena such as these do not per se constitute
a theory, rather they are a crucial component of and target for systematic
explanation. At the same time, they constitute hard-won contributions to
scientific knowledge, which are developed and supported through the production and analysis of data.

The proposal that theories predict and explain phenomena, rather than
data, was a major intervention in the debates over theory-ladenness and
theory-data relations that I briefly sketched above. Predictably, given the
theory-centric bias within the philosophy of science, this generated three
decades of debate over the ontological status of phenomena vis-à-vis theories, and the role of phenomena as evidence, while relatively little attention
was devoted to the implications of this view for conceptualizing data.[40] Like
Hacking, Bogen and Woodward emphasized the "raw" nature of data as
immediate outputs of measurement and/or experimentation. Woodward in
particular contributed a definition that highlights this feature: "Data are
the individual outcomes of a measurement or detection process, which may
involve instruments or unaided human perception."[41] As a consequence of
this definition, Bogen and Woodward argue that data are "idiosyncratic to
particular experimental contexts, and typically cannot occur outside of
those contexts," while claims about phenomena have "stable, repeatable
characteristics" and can thus "occur in a wide variety of different situations
or contexts."[42] In other words, data carry information about what the world
is like, but such information is expressed in ways that can only be properly
understood and interpreted by scientists who are familiar with the setting

in which data are acquired. Knowledge about phenomena need to be freed from its embedding in data (and thus in the local practices through which data are obtained) in order to be shared among scientists irrespective of their familiarity with the means through which it has been acquired. Such freedom comes through the formulation of claims about phenomena. Data help scientists infer and validate those claims, yet it is ultimately the claims about phenomena that travel and are used as evidence for general theories. Hence, for Bogen and Woodward, data are local evidence for nonlocal claims.

The idea that data and claims about phenomena have intrinsic degrees of locality, regardless of the ways in which they are used, stands in stark contrast to my relational framework and proves problematic on empirical grounds. It does not help with cases in which data are manipulated and/or abstracted and yet are still used as "raw data" by researchers. It also does not account for cases where data travel beyond their original site of production and are adopted as evidence for new claims, nor does it account for the extent to which the traveling of claims about phenomena depends on shared understanding across epistemic cultures. It is true that data published in a research article are selected as evidence for one specific claim about phenomena, and that what readers are required to take away from such a paper is not the data themselves but rather the empirical interpretation of those data that is provided by the authors in the form of a claim. However, this is precisely the situation that data-centric research lobbies such as the Open Data movement regard as problematic, because it effectively prevents data from traveling beyond their original site of production. When viewed in this way, locality becomes a consequence of a specific regime of data dissemination, rather than a characteristic of data themselves; and the development of alternatives to such regime, such as that embodied by model organism databases, can affect the nature of the ties between data and specific research situations.

My analysis of data packaging shows that journeys via databases expand the evidential value of data in several ways. They make data accessible to other research contexts, and therefore potentially reusable as evidence for new claims, and associate data with a broader range of phenomena than the one to which they were associated in the production context. This brings me to contest the idea that data are intrinsically local, by pointing out that data can be made nonlocal through the use of appropriate packaging processes. Data disseminated through databases are linked to information about their provenance, but they can be consulted independently of that information. As I detailed in chapter 1, this is a way to decontextualize data and transform them into nonlocal entities, since the separation of data from information about their provenance allows researchers to judge the data's potential relevance to their research. By contrast, judging the reliability of data within a new research context requires database users to access information about how data were originally produced and assess it through their own (local) criteria for what counts as reliable evidence, based on the expertise that they have acquired through their professional experience. Thus data judged to be reliable are recontextualized and become local once again: what changes is the research situation in which they are appropriated.[43]

Bogen and Woodward's account also maintains that claims about phenomena are intrinsically nonlocal, or anyhow less local than data themselves. This position also loses plausibility when taking data journeys into account. My analysis of packaging shows how scientists' interpretation of the terms used in claims about phenomena is always situated by their specific background knowledge and skills. As I will illustrate when examining data labels in chapter 5, the classification and definition of phenomena crucially depends on the interests and expertise of the scientists who investigate them. This is what makes it hard for curators to develop nonlocal labels for the phenomena for which data may function as evidence, since the development of nonlocal labels require substantial (and not always successful) mediation between the local cultures of research at the bench. Like data, claims about phenomena only acquire nonlocal value through apposite packaging of the terms used to refer to phenomena and the efforts made by different epistemic cultures to position and interpret those claims according to their own expertise, commitments, and background assumptions. The resulting nonlocality of claims is as much a scientific achievement as the nonlocality of data.[44]

One of the reasons why Bogen and Woodward appeal to the intrinsic nonlocality of claims about phenomena is to defend the idea that "facts about phenomena are natural candidates for systematic explanation in a way in which facts about data are not."[45] I agree with this intuition, which is compatible with the framework I provided here. What I wish to contest is the idea that this difference can be accounted for as a question of locality. What marks the difference between data and claims about phenomena is the way in which they are used at specific stages of inquiry, which is in turn partly determined by the specific format of these entities and the ways in which they travel. Claims about phenomena are privileged candidates for systematic explanations because they are propositions, rather than dots on a slide, photographs, or numbers generated by a machine. Their linguistic format places them at a relatively high level of abstraction from the entities and processes that they are taken to document and enables them to be integrated into formal structures such as theories or explanations. At the same time, propositions can also be used as data; this happens most often in the case of observation statements used to corroborate a claim about the phenomena being observed (e.g., the statements "the frog saw the snake" and "the frog jumped into the lake" can be used as evidence for the claim "frogs take refuge when spotting a predator"), although observation statements can also function as claims about phenomena depending on the situation at hand (e.g., when attempting to evaluate the evidential value of a series of photographs of a frog's movement in the field, which can be considered as evidence for the claim "the frog jumped into the lake"). Furthermore, propositions functioning as claims about phenomena can typically be used on their own and in association with other claims.[46] Data, by contrast, tend to travel and be used in groups. Indeed, cases where one piece of datum suffices as evidence for a claim are extremely rare. To become evidentially significant, data find strength in numbers. The more data are grouped together, the stronger their evidential weight, as large numbers of data points may increase the likelihood of finding significant correlations, and their accuracy. This is particularly true of data in numerical forms, which are most easily subjected to statistical analysis and mathematical modeling. The gregarious nature of data, and their susceptibility to statistics and other sophisticated techniques of analysis and interpretation, contribute to explaining why the size of datasets is often regarded as a significant factor in determining their reliability and potential evidential value—the current emphasis on the power of big data being a case in point.[47]

3.4 Packaging and Modeling

I now consider in more detail the idea that the evidential value of data depends on the ways in which they are arranged, presented, and visualized. The analysis of datasets found in databases, often referred to as "data mining," involves devising ways to visualize results. Visualizations can reveal patterns that would not be spotted unless data are adequately displayed; they also offer a potential solution to the problems posed by the quantity and quality of the data to be analyzed, particularly in cases of very large or highly diverse datasets.[48] Philosophers typically interpret visualization processes as forms of modeling that provide an essential bridge between data production and data interpretation. A seminal contribution is Patrick Suppes's work on "models of data"—that is, models "designed to incorporate all the information about the experiment which can be used in statistical tests of the adequacy of the theory."[49] The emphasis in this account is on the crucial role of statistics in helping scientists to fit scattered data points into a significant pattern, which can be used to test theoretical predictions or, in Bogen and Woodward's terms, to corroborate claims about phenomena.

Suppes focuses on the ways in which statistics helps scientists to abstract and simplify data away from the "bewildering complexity" of experimental situations. He thus separates models of data from models of experiment, which describe choices of parameters and setup, and practical issues such as sampling, measurement conditions, and instrumentation, which he views as crucial "pragmatic aspects" of experimentation that precede modeling activities.[50]

In Suppes's picture, concerns and decisions involved in data production are only relevant to modeling as long as they affect statistical analysis; and concerns around data dissemination are not mentioned at all.[51] His notion of data is never explicitly defined and seems broadly compatible with Hacking's view of data as immediate experimental outputs. In this sense, his analysis deals only with a specific subset of what I regard as data—those subjected to specific kinds of statistical manipulation. Nevertheless, and in parallel with Rheinberger's questioning of the physical manipulation of experimental traces, his concern with the epistemic status of data processing procedures raises a crucial question: What does it mean for data to be "abstracted away" from their original format and production site? This issue is particularly significant within a relational framework in which data are defined by their function as prospective evidence. Models of data can and do serve as evidence for claims, which seems to assign them the same function of data themselves; at the same time, these objects have undergone modifications that make them significantly different from the ones originally created by experimenters. Roman Frigg and Stephen Hartmann stress this aspect in their own definition of models of data as a "corrected, rectified, regimented and in many instances idealized version of the data we gain from immediate observation, the so-called raw data."[52] What is it that sanctions the difference between data and models and also between modeling processes such as those described by Suppes and the activities of data packaging that I have discussed so far?

This question is particularly difficult to address in light of the wide diversity in the formats, sources, and uses of both data and models within and across areas of research. Paul Edwards has provided a particularly detailed study of the relationship between data and models in climate science, which constitutes a useful starting point for my own reflections on data modeling in the life sciences. Edwards presents data handling and modeling as interrelated activities and repeatedly stresses that data are theory-laden as much as models are data-laden.[53] His study focuses on data assimilation, a process that he describes as requiring a "profound integration of data and models,"[54] as key to the packaging of data for travel. Thus Edwards recognizes the difference between data and models but stresses the importance of modeling practices as principal means for data interpretation, and in fact as the only feasible form of data handling in science. If we follow this view, the data packaging processes that I have discussed up to this point should be conceptualized as modeling activities, and there is no point in singling out data dissemination practices as a significant part of research.

However, Edwards himself offers a reason to reject this conclusion when he discusses what he calls "data wars" in climatology, which involve two distinct ways of handling data. One is championed within weather forecasting, where the idea of "raw data" is not highly valued ("original sensor data may or may not be stored; usually they are never used again"[55]) and scientists tend to work with models of data built through statistical tools, much in the way described by Suppes. The other is practiced in climate science, where interpretation happens at different points in time in relation to a variety of research questions, and is based on data collected across vast periods and diverse geographies. Within this latter camp, scientists put much effort in preserving what they view as "raw data" and documenting the ways in which their format is modified in order to make them travel—an approach similar to the biological cases I have been discussing. Both in the climate and life sciences, the packaging of data is not solely or even always focused on statistical analysis; it also involves work on the medium, format, and order of data, as well as the ways in which they are combined with other datasets or selected/eliminated in order to fit a certain kind of vehicle (such as a database). These activities are primarily focused not on building links between the ways in which data are visualized and a given target system but rather on securing reliable sources of evidence for knowledge claims.

This is not equivalent to modeling because, while it affects the ways in which data are interpreted, it is not meant as an act of interpretation and does not necessarily teach scientists anything about the target systems that they are investigating. The main concern for researchers and technicians engaged in data packaging is to monitor and possibly improve the procedures through which data are generated and made useable as evidence. Data packaging is subject to specific constraints and challenges such as data formatting (e.g., how to translate data from one medium to others, so that it can travel through vehicles as diverse as notebooks, journal articles, and online databases?) and the choice of metadata (which information about data provenance best informs future data interpretation, and how can it be efficiently linked the relevant data?). These constraints may well affect subsequent attempts to interpret the data, but this is not always intended or monitored by whoever is responsible for such manipulations, whose primary goal is to make it possible for data to travel outside of their production context. As I discuss in the next chapter, the curators of model organism databases explicitly recognize the importance of data packaging decisions for prospective data interpretation, but they are not able to predict exactly what impact these choices could have on the evidential value of data—that depends on the specific research settings within which data are eventually adopted and used.

The study of data journeys illustrates that making data portable may involve but does not require, nor does it aim at, the interpretation of data as representations of specific phenomena. Models, by contrast, are typically manipulated with the explicit intent to represent some feature of the world.[56] This characterization of models is not tied to a specific interpretation of the notion of representation, nor does it make assumptions about the nature of the phenomena in question, which may be conceptualized as fictional or actual, constructed or found in nature, and preexisting investigation or resulting from the process of inquiry itself.[57] All that matters for my purposes is the recognition that modeling involves the creation of tools that can bridge between the imagination of scientists and the world, in ways conductive to developing new knowledge.[58] The modeling of data can thus be described as attempts to learn about the world by inferring information about a phenomenon from one or more specific visualizations of data. This is indeed a process of abstraction, insofar as researchers use data to identify, describe, and/or explain specific features of a phenomenon of interest.[59]

Both data and models are objects that researchers continuously manipulate and intervene on in their quest for knowledge. I have argued that the difference between them lies in the goals ascribed to those manipulations and the constraints under which they take place. The same object can function as data or as model, depending on whether it is intended to function as evidence for claims or to teach researchers something about the world. Modeling is the process by which data are assigned evidential value. Data packaging is the process by which data become amenable to analysis in the first place. This reading is compatible with Suppes's remarks on models of data but places due emphasis on the crucial epistemological role played by intuitive considerations underlying the production as well as the dissemination of data, which provide the conditions under which data are modeled and therefore shape the procedures and outcomes of modeling efforts. Choosing which interpretation best fits a given dataset depends not only on the statistical principles used to analyze them but also on the features of the data under consideration, the ways in which they are ordered and visualized, and the interests and expertise of the scientists involved.

In closing, let me come back to the two apparently contradictory perceptions of data as "given" and "made" that, as I noted at the start of this chapter, have long plagued philosophical and scientific discussions of data analysis. Christine Borgman has remarked that data "may exist only in the eye of the beholder: the recognition that an observation, artifact, or record constitutes data is itself a scholarly act."[60] This perspective clashes with the expectation, often felt within the sciences and voiced by Borgman herself, to specify what data are as objective, context-independent units. And yet researchers do not need to conceptualize data as objective and context-independent units in order to foster their evidential value. On the contrary, the packaging strategies developed by database curators explicitly acknowledge the subjective, context-dependent nature of data as a starting point to foster their travel and evidential value. Within model organism biology, attempts to collect and disseminate data are often based on the assumption that as many data as possible should be collected, and that they should be made available for the long term as raw materials for further analysis. At the same time, these data are widely accepted to be constructs, whose value and quality are affected by the conditions under which they are obtained. Whether data found in databases can have evidential value beyond the species or even the organism on which they are obtained needs to be evaluated in every single case of travel. By packaging data in ways that include information about their provenance and subsequent journeys, biologists manage to reconcile the wish to reuse data across research contexts with the awareness of how the significance of any one dataset may vary dramatically depending on the species and biological phenomenon to which it is brought to bear, as well as the transformations that it underwent in its travels. The procedures through which data are produced and disseminated make them unique sources of information; and the ways in which those procedures are documented determine how data will be interpreted across contexts. Thanks to packaging processes that enable researchers to recontextualize data in their own terms, with critical awareness of the manipulations that they have undergone during their journeys, the man-made quality of data does not compromise their ability to serve as reliable evidence for claims about phenomena.

4

What Counts as Experiment?

I now take a closer look at the links between data dissemination and data production, by addressing the epistemic relation between experimentation carried out in vivo and data curation and analysis conducted in silico. Little philosophical reflection has addressed the problem of what makes it possible for researchers to interpret data available online and assess their evidential value. I will argue that in experimental biology, this process requires some degree of familiarity with the target system that data are taken to document and the experimental conditions under which that system is studied. This does not mean that data users need to be acquainted with the specific circumstances under which data have been generated, but rather that they need to have some experience in the experimental manipulation of actual organisms, which they can use as a comparator to evaluate information about data provenance.

My entry point into this discussion is an analysis of the ways in which the curators of model organism databases handle information about the procedures, protocols, skills, teamwork, and situated decision making that are involved in the creation of data. As I illustrated in chapter 1, curators are well aware that metadata that provide information on the life history of data are crucial to assessing their quality, reliability, and evidential value. To capture those histories, curators put considerable efforts in the development of standardized descriptions of the techniques, assumptions, methods, materials, and conditions under which data are generated. The first part of this chapter reviews some of these attempts and argues that curators engaged in the development of metadata are effectively attempting to formalize *embodied* knowledge that plays a key role in validating the claims about phenomena (*propositional* knowledge) for which data serve as evidence.

I then turn to the problems that plague these efforts. While it is indeed possible to capture embodied knowledge through tools such as proposition and images, such codification is not sufficient to communicate (and thus replace) embodied knowledge gained through scientific practice. Access to metadata needs to be accompanied by concrete experience in the material interactions being described, which greatly improves researchers' ability to learn from others' behaviors and findings. Put more crudely, reading about embodied knowledge does not necessarily teach how to enact such knowledge. A researcher who accesses metadata from a database will hardly know how to perform the experiments being described and/or how to assess the evidential value of the resulting data—unless she already has some experience in performing similar practices of data production.

These observations do not condemn efforts to codify and communicate embodied knowledge as intrinsically flawed or useless. On the contrary, I regard them as crucial to improving the accountability of research and stimulating methodological debates around the study of living organisms. One of the functions of current efforts to develop metadata is to offer a platform for the expression and critical discussion of the features, validity, and replicability of activities involved in experimentation. This encourages researchers to consider and reassess their own way of working, by comparing it with the procedures favored by other labs. It also helps to evaluate the compatibility of assumptions and decisions taken by experimenters with those taken by the many other experts involved in data journeys, including for instance statisticians, computer scientists, and software engineers. The explicit discussion of the varieties of embodied knowledge involved in data interpretation points to the peculiarly distributed nature of data-centric biological reasoning, which is best viewed as a collective rather than an individual achievement.

I conclude that scrutinizing the role of embodied knowledge in data journeys demonstrates how physical interventions on organisms, and the social interactions that make those possible, retain epistemic value vis-à-vis their representation within databases. While there is no doubt that research grounded on database mining is playing an increasingly important role in complementing and supporting experimental work, the interplay be-

tween these two approaches remains crucial to obtaining valid and significant knowledge about the natural world. The communication of embodied
knowledge through metadata should thus not be conceptualized as means to
do away with experiments altogether, since the interpretation and reuse of
data found online still depends on researchers' experience in manipulating
organisms under controlled conditions; nor does it guarantee experimental
replicability, since the ways in which researchers with different backgrounds
will understand and act on such information may vary widely depending
on their experiences at the bench. Consequently, I contest the idea that discovery through the analysis of large datasets can ever be fully automated
and/or used as a substitute for experimental intervention in vivo.

4.1 Capturing Embodied Knowledge

Embodied knowledge is that required to physically handle data and use
them as evidence within a specific research context. More broadly, it can be
defined as the awareness of how to act and reason in order to pursue scientific research, which is essential for biologists to intervene in the world, improve their control over the systems that they study, and handle the data and
models thus produced.[1] This notion parallels what Gilbert Ryle calls "knowing how": the series of actions and skills involved in the material realization
of an experiment.[2] Embodied knowledge is expressed, for instance, through
the procedures and protocols that allow scientists to intervene on the entities and processes of interest; the ability to implement those procedures and
modify them according to the specific context of research; the skills involved
in handling instruments, models, and organic samples; the perception, often based on the scientists' experience in interacting within a specific space
(such as a laboratory), of how to move and position oneself with respect to
the models and materials under study; and the development of methods to
facilitate the replication of experimental results.

In this section, I show how the codification of embodied knowledge in biological databases runs counter to the philosophical vision of such knowledge
as "tacit" and "personal," as famously advocated in the writings of Michael
Polanyi. Polanyi depicted the ability to intervene effectively in the world in
order to develop or apply propositional knowledge as an "unspecifiable art,"
which can only be learned through practice or imitation and which is rarely
self-conscious, as it needs to be integrated with the performer's thoughts
and actions without swaying her attention from her main goals.[3] This is
why Polanyi refers to this knowledge as tacit: it is an ensemble of abilities
that inform scientists' performances (actions as well as reasoning), without

being directly acknowledged by the individuals who possess them. Polanyi furthers this interpretation by distinguishing between two ways in which scientists can be self-aware: "focal" awareness, which is tied to the intentional pursuit of "knowing that," and "subsidiary" awareness, which concerns the scientists' perception of the movements, interactions with objects, and other actions that are involved by their pursuit. In this view, focal and subsidiary awareness are mutually exclusive: one cannot have one without temporarily renouncing the other. Switching from "knowing that" to "knowing how" and back is, to Polanyi, a matter of switching between these two ways of being self-aware. In fact, his characterization of know-how as "subsidiary awareness" relegates this type of knowledge to a secondary position with respect to "knowing that."[4] The acquisition of theoretical content constitutes, according to Polanyi, the central focus of scientists' attention. Tacit knowledge makes it possible for scientists to acquire such theoretical content, but it is not valuable in itself: it is a means to a (theoretical) end.

While drawing inspiration from Polanyi's emphasis on tacit expertise, I do not think that his hierarchical ordering of "knowing how" as subsidiary to "knowing that" does justice to the intertwining of propositional and embodied knowledge that characterizes biological research. Within most of biology, the development of skills, procedures, and tools appropriate to researching specific issues is valued as highly as—or, sometimes, even more highly than—the achievement of theories or data that it enables. The evolution of tools and procedures is crucial to conceptual developments, and biologists typically acknowledge this to be the case. Ryle's view on embodied knowledge is thus more congenial to my analysis, as it does not assume either type of knowledge to be dominant over the other. This is due to an underlying difference between his outlook and Polanyi's. While the latter thinks of the distinction as descriptive, thus presenting the two types of knowledge as separate in practice, Ryle treats the distinction as an analytic tool and refers to its descriptive interpretation as an "intellectualist legend": "When we describe a performance as intelligent, this does not entail the double operation of considering and executing."[5] Ryle defines embodied knowledge as involving both intentional actions and unconsciously acquired habits. In fact, he prefers an intelligent agent to avoid as much as possible the enacting of habits: "It is of the essence of merely habitual practices that one performance is a replica of its predecessors. It is of the essence of intelligent practices that one performance is modified by its predecessors. The agent is still learning."[6] And further, "To be intelligent is not merely to satisfy criteria, but to apply them; to regulate one's actions and not merely to be well regulated. A person's performance is described as careful or skillful, if in his operations he is ready to detect and correct lapses, to repeat and improve upon success, to profit from the examples of others and so forth. He applies criteria in performing critically, that is, in trying to get things right."[7]

Ryle does not specifically apply his account to the case of scientific knowledge. Nevertheless, his characterization of agents "trying to get things right" by iteratively improving their performance can be neatly applied to biologists' attempts to codify embodied knowledge, which enable them to work together on improving existing techniques, methods, and procedures. Embodied knowledge is neither tacit nor personal in data journeys. Rather, its expression and formalization into databases, guidelines, protocols, and related standards is crucial to the circulation and reuse of data in the first place. For data to travel, embodied knowledge needs to be made public and thus be codified into descriptions, instruments, and materials that can be shared among individuals and groups. Indeed, we have already seen how packaging data for travel involves the development of metadata documenting where specific datasets originally come from; how they were collected; and how they were formatted, annotated, and visualized upon entry into the database. This involves making decisions about which aspects of the embodied knowledge involved in generating data are most relevant to situating them in a new research context, integrating them with other data, and interpreting their significance, as well as finding ways to formalize such knowledge that can be intelligible to researchers who do not have direct experience of the techniques, materials, and organisms involved. In what follows, I review some of the efforts made by the curators of community databases in biology toward accomplishing these two tasks.

One obvious way to capture embodied knowledge is through text—that is, descriptions of how activities of data production have actually been performed. This is not a new idea: lab books, protocols, and methodological guidelines have long been employed by experimenters to communicate embodied knowledge among each other.[8] The trouble with these descriptions is that they tend to pay homage to the peculiarities of local research environments and traditions and refrain from employing standard terms or formats. They are also somewhat erratic, as there are no general rules for what they should contain, and their consultation will rarely suffice for a group other than the original to be able to replicate the experiment. Furthermore, they are not routinely circulated beyond the boundaries of the lab in which the related activities are performed, as researchers tend to share these materials only upon request or in cases of suspected misconduct. Indeed, despite holding ground as a well-recognized desideratum in scientific inquiry, the expectation that valid experiments should be replicable does not seem to have imposed strong constraints on which experiments are ultimately judged to produce reliable scientific results; and, while it certainly has encouraged scientists to carefully record the circumstances in which data are obtained, it has not fostered the widespread adoption of common standards for the articulation of embodied knowledge.

Another increasingly popular means to capture embodied knowledge is the use of nontextual media such as graphs, podcasts, and videos. The latter in particular purport to provide a more faithful and instructive rendition of experimental circumstances than standardized texts. A notable example is *JoVE* (*Journal of Visualized Experiments*), a peer-reviewed video journal established in 2006 to publish scientific research in a visual format. *JoVE* presents itself as aiming to "overcome two of the biggest challenges facing the scientific research community today: poor reproducibility and the time and labor-intensive nature of learning new experimental techniques."[9] This self-portrayal parallels Polanyi's picture of embodied knowledge as acquired through painstaking processes of imitation and apprenticeship and thus emphasizes the ability to view how experiments are conducted in real time as the closest experience to witnessing such performance in the lab. I certainly do not wish to disqualify the importance of such efforts: the immense popularity of "how to" DIY videos on social media such as YouTube is a glaring reminder of the effectiveness of videos as means to communicate embodied knowledge both within and beyond science, and the availability of well-produced short films can spare biologists the need to physically travel between labs to learn a new method. At the same time, videos retain many of the problems associated with traditional lab books and individual visits. They are time-consuming to watch, highly idiosyncratic, hard to compare with each other, and arguably still in need of supplementary information in order to work as a basis for experimental replication.

To make data travel through databases, many database curators have started to work toward improving existing ways of capturing experimental procedures, including both textual and nontextual forms, so that accounts of embodied knowledge can be more easily developed, disseminated, compared, and understood. A prominent effort associated to the building of model organism databases has been Minimal Information on Biological and Biomedical Investigations (MIBBI), a system for the standardized textual descriptions of the most basic embodied knowledge involved in biological experiments, which data donors, curators and users can employ to develop metadata.[10] Another example is provided by the Plant Ontology (PO), which proposes both textual and graphical standards to depict the morphological features of the plants used in model organism experiments—starting from the traits of the thousands of plant mutants stored in *Arabidopsis* stock centers.[11] In this case, what curators are trying to capture is not the activities performed by researchers but rather the characteristics of the specimens used, which play a significant role in shaping the skills, instruments, methods, and activities that researchers choose to exercise in any given experiment.

MIBBI and PO should not be viewed as simple-minded attempts to homogenize experimental procedures across laboratories. Both initiatives take as their starting point the recognition that each experimental setting is unique, due to extreme complexity of the parameters involved. Their attempt at standardization is based not on the simplistic proposal that that "one size may fit all" but rather on a sophisticated view of the role and functioning of embodied knowledge in biology, which could be summarized as involving three core ideas.

First, some features of an experiment matter more than others when it comes to assessing the quality and significance of the results obtained. For instance, it is impossible for a biologist to determine the significance of a given dataset in the absence of information about what materials that dataset was taken from (such as which model organism and, if known, which mutant strain, phenotype, and genotype).

Second, these features can be singled out as pertaining to the same "type" across several experimental settings. Any researcher wishing to interpret biological data found online will need information about what organism they were taken from, with which instrument(s), who carried out the experiment, where, and when. So metadata will need to include categories such as "organism," "instruments," "authors," "location of original experiment," and "time of original experiment."

And third, at least some of these features can be explicitly described through texts, graphs, or other media, thus enabling their communication across large networks of individuals without making it necessary for those individuals to witness how these procedures are realized in the lab.

The success of this vision of embodied knowledge is illustrated by the staggering speed with which interest and investment in the development of metadata has developed since the turn of the millennium. By the start of 2014, MIBBI and PO had joined another 544 initiatives under the general umbrella of BioSharing, a web portal devoted to assembling "community developed standards in the life sciences . . . to make data along with experimental details available in a standardized manner."[12] Initiatives within BioSharing target three main types of standard descriptions: *reporting guidelines* on various experimental processes (including sample processing, magnetic resonance, and randomized controlled trials), *exchange formats* of relevance to different types of data (e.g., markup language to be used with images, clinical results, genome sequences), and *labels* proposed as standard terminology to describe knowledge within specific fields (such as anatomy, chemistry, and systems biology) and the anatomical characteristics of various organisms (e.g., plants and vertebrates, and of course specific models like *C. elegans, Dictystelium discoideum, Drosophila,* Xenopus, and zebrafish).

These means to codify knowledge are used as metadata accompanying data in their journeys through databases and across research contexts. Their role in data journeys reinforces the impression that, in Evelyn Fox Keller's well-known formulation, determining the scientific significance of data requires "a feeling for" the material conditions under which phenomena are investigated and the ways in which specific samples—in my case, parts and specimens of model organisms—instantiate the phenomena in question.[13] Keller's expression has been critiqued for its lack of precision, a vagueness that plagues arguments pointing to the importance of embodied, nonpropositional knowledge in scientific research. Metadata become interesting from the epistemological viewpoint insofar as they constitute an explicit attempt to articulate the so-called tacit dimensions of scientific knowledge. By supplying as much information as possible about how data are produced, metadata become a tool to dispel at least some of the vagueness associated with experimental activities and skills and foster the extent to which such knowledge is reported, taught, and assessed across individuals and groups.

4.2 *When Standards Are Not Enough*

Recognizing the role of metadata in communicating embodied knowledge does not necessarily involve challenging the idea that biological experimentation is a localized, situated affair, each instance of which clusters and reconfigures diverse ensembles of skills, assumptions, materials, environmental conditions, and goals; nor does it mean rejecting the crucial role played by training at the bench, on actual organisms, in obtaining the embodied knowledge involved in experimental biology. In other words, I do not believe that the implementation of metadata necessarily encourages the standardization and homogenization of experimental procedures in biology, nor do I think that it enables researchers without bench experience to interpret data on model organisms as reliably and accurately as those who handle those organisms daily. To illustrate how these ideas can be reconciled, I now analyze some of the problems surrounding the use of metadata in model organism biology, which illustrate both the effectiveness of such codification in encouraging debate around data production and interpretation procedures and the extent to which the intelligibility of these formalizations depends on database users' ability to compare such information with their own experiences at the bench.

Attempts to codify embodied knowledge provoke researchers into comparing and debating their experimental procedures; they also raise the question of whether and how experimental materials and procedures can be made homogeneous across different laboratories and research cultures, and to what effect. Even within relatively cohesive communities such as those working on specific model organisms, research groups tend to disagree on what elements should be prioritized when describing the provenance of data. Furthermore, within any one project, experimental protocols and procedures are in constant flux, as biologists adapt them to the ever-changing demands and local conditions of their research. This makes it even more difficult to settle on fixed categories for the description of experimental activities. Edwards and his collaborators have discussed debates over the formulation and use of metadata as cases of "data frictions."[14] In what follows, I examine two examples of such frictions specific to experimental research in the life sciences, which exemplify both the difficulties in formulating metadata and the usefulness of such efforts in getting biologists to explicitly tackle and discuss discrepancies in experimental practices.

4.2.1 Problem Case 1: Identifying Experimental Materials. The first example concerns the identification of the very materials on which experiments are performed—which, in our case, consist of organic specimens (parts or whole exemplars of organisms). To understand this example, it is important to note that experimentation on model organisms often aims to explore the reasons for the morphological, genetic, and/or physiological differences among variants of the same species. Because of this, knowing which strain of the organism in question was used to generate a given dataset constitutes a key piece of information for any researcher wishing to reuse those data. It therefore seems reasonable to assume that information about organic specimens should always accompany data when they travel from one context to another. This constitutes an ideal case of "minimal" information about experimental practice in biology and is recognized as such by several BioSharing systems, including MIBBI.

Nevertheless, capturing information about these experimental materials proves to be difficult as well as contentious in practice, since researchers working in different experimental settings harbor divergent views on how to report information on the materials that they use. For instance, some of the database curators that I interviewed pointed to differences among clinical and nonclinical settings. In their experience, clinical researchers often fail to specify which organic samples they are working on—for instance, whether they have mixed samples coming from different organisms (e.g., human cell cultures and animal RNA probes), or whether they have worked on rat or mouse cells. This clashes with the attitude preferred by researchers working on model organisms in nonclinical settings, for whom such information is vital to the interpretation of the resulting data. As a result of researchers' inability to specify this type of metadata, data acquired in clinical settings often fails to be included in model organism databases altogether. In the words of one curator, "When people publish, they . . . a lot of times don't tell you what protein they're working on, whether it's mouse or human. They'll tell you the protein name, but that could be 99% identical between human and mouse and they won't tell you which species it's from. And so in that case we can't annotate, we don't know exactly the species."[15]

Another reason for the difficulties surrounding this type of metadata lies in the large variety of strains of model organisms that are utilized in research. On the one hand, most of such organisms display high variability across environments, so researchers who collect them from diverse locations end up with many different types of specimen whose features need to be systematically compared. On the other hand, experimental processes themselves, by radically modifying the lifestyle of organisms and of course through genetic engineering, end up producing large quantities of mutants. Metadata are supposed to capture the specificity of the strains used in any given experiment, but this is hard to do in a situation where hundreds of different strains are used and modified daily within thousands of laboratories around the world. The descriptions of those strains tend to be couched in nonhomogeneous terminologies and languages, making it difficult to compare experimental accounts and extract common information. Agreeing on classifications and descriptions that may facilitate the accurate tagging of organisms is an immense logistical exercise, requiring considerable resources and cooperation across dispersed research communities.

The development of metadata that adequately capture information on experimental specimens has been most successful in cases of relatively well-organized communities focusing on organisms that are easy to store and transport and that have no significant economic value—such as *Arabidopsis* (whose mutants can be circulated and stored in the form of microscopic seeds), *E. coli* (in which case, whole mutant colonies can be easily contained in small Petri dishes), and *C. elegans* and zebrafish (whose mutants are also relatively easy to keep and disseminate, due to their size and relative robustness to environmental changes). In these cases, the management of strains is overseen by centralized stock centers, whose task is to provide reliable ways of identifying the mutants and accessing them for future experiments.[16] The work of the technicians and managers working in these stock centers has become a central component of data journeys, as it guarantees the provision of accurate metadata about the materials used to generate each dataset and enables users to order such materials if they wish to replicate or continue that line of work. The effectiveness of these metadata is enhanced by the fact that most researchers who access them will have experience in manipulating organisms of a similar type, even if not exactly the same strains. This improves the ability of database users to assess the provenance of data and takes pressure off curators, who can rely on experimenters to provide feedback whenever a classification does not match their experience.[17]

The vast majority of species used within experimental biology is not stocked and managed in this centralized way. This makes the provision of accurate metadata on organic materials exceedingly difficult, as well as heightening the risk that database users might not be able to relate information provided through metadata to their own experiences in the lab. A notable case is that of mice, whose many strains constitute a primary source of materials for biomedical research and yet whose classification and collection tends to be highly fragmented. Historians and social scientists have highlighted reasons for this state of affairs, including the immense popularity of mice within private facilities, the variation among national laws regulating their use in research, the high costs involved in their transport and maintenance, and the history underlying the commercialization and licensing of some of these strains, as in the iconic case of the OncoMouse.[18] Another problematic case is that of microbes, whose rapid evolution and mutability in the lab makes them difficult to describe through fixed categories.[19] Small research communities working on relatively unpopular organisms tend to fare better when circulating data among their members, because they can rely on personal contact and acquaintance with each other's work in ways that much larger communities cannot; however, they are at a disadvantage when attempting to make their data travel beyond their restricted network, as required when contributing to cross-species databases.

This example illustrates how difficult it is to provide even the most elementary information about experimental procedures in a way that is universally unambiguous and reliable. At the same time, the requirement to make model organism data travel, particularly the formalization of metadata within databases, has played an important role in bringing these difficulties to the attention of the scientific community. It is not a coincidence that the rise of model organism databases has coincided with a renewed interest to rally against the lack of standardization in collections of materials and related difficulties in accessing and comparing model organisms. An example is provided by CASIMIR, a project funded by the European Commission between 2007 and 2011 to improve the CoordinAtion and Sustainability of International Mouse Informatics Resources (hence the acronym). This project was primarily aimed at providing reliable databases to foster the use of mice as models for human diseases; and yet many of its initiatives ended up focusing on the identification and accessibility of mice strains, as well as the metadata used to describe specimens.[20]

4.2.2 Problem Case 2: Describing Laboratory Environments. The second example concerns the descriptions of the environmental conditions under which data are generated. An interesting case is the dissemination of microarray data about gene expression, which constitutes one of the most popular and widely disseminated "omics" data in contemporary biology (see figure 3). Microarrays are produced in a digital format through largely automated experiments, which are conducted with the help of highly standardized machines. They are hailed as big data that are intrinsically portable and easy to integrate regardless of their provenance. And yet many researchers view microarray data as potentially unreliable and difficult to replicate across labs, due to their susceptibility to environmental conditions in which the experiments are carried out. Variations due to factors such as temperature and humidity affect the quality of the data produced, and yet it is hard for researchers to agree on metadata that adequately capture these variations— not least because there is little consensus on which environmental factors have the strongest impact.[21] These difficulties are reflected in the work of database curators, who are well aware that their users do not harbor a high level of trust in microarray results acquired by other researchers in the absence of adequate information about experimental conditions.[22] In 2000, a group of biologists set out to address this issue by developing Minimal Information About a Microarray Experiment (MIAME), so as to streamline the process of agreeing on, and implementing, adequate metadata.[23] Attempts such as this attracted a lot of controversy and contributed to highlighting the problems involved in replicating microarray experiments.[24] For instance, in 2009 *Nature Genetics* published a study that evaluated the replicability of eighteen datasets obtained through microarray experiment and found that ten could not be reproduced on the basis of the information provided, from which it inferred that "repeatability of published microarray studies is apparently limited. More strict publication rules enforcing public data availability and explicit description of data processing and analysis should be considered."[25] These conclusions generated a debate that is still ongoing at the time of writing, including studies claiming that the use of richer models of the process by which microarray data are constructed leads to better and more reproducible results of even existing microarray data.[26]

When we consider data obtained through procedures that are less highly standardized than microarray experiments, these issues become even more substantial. Take the case of metabolomic experiments, which attempt to capture as much of the chemical composition of a tissue extract as possible and to compare it among samples. This is difficult to achieve because, unlike nucleic acids and proteins, the diversity of small molecules in a tissue sample means that no one extraction and analytical technique can detect everything. Added to this, most organisms do not have a reference metabolome against which comparisons can be made and no exact figure can be put on the existing number of small molecules. Metabolomics data are also diverse in the collection techniques used and their degree of complexity, and even data produced through the most popular of such techniques (mass spectroscopy) turn out to be difficult to interpret, because features detected in these experiments cannot be identified or verified without standard compounds or further purification and detailed analysis.[27] These difficulties in adequately curating and interpreting metabolomics data account for the scarcity of metabolomics data currently available in public databases. Efforts to improve this situation, while also taking account of the specificity of the circumstances in which metabolomic data are collected, are increasing. Key to these efforts is the engagement of users in actively assessing and complementing the embodied knowledge captured by the databases. For instance, Metabolome Express (http://www.metabolome-express.org) includes tools for extraction of information from raw data files as well the ability to carry out comparative analyses. Databases like the Arabidopsis Metabolomics Consortium (http://plantmetabolomics.vrac.iastate.edu/ver2/index.php) and the Platform for Riken Metabolomics (PRIMe; http://prime.psc.riken.jp) are even more project specific, requiring a high level of experience in metabolomic experiments to be able to analyze the data therein. To help researchers who do not possess as much relevant embodied knowledge, there are a number of databases storing nuclear magnetic resonance (NMR) spectra and mass spectroscopy (MS) spectra that can be used for aiding compound identification, for example: Golm Metabolite Database (http://gmd.mpimp -golm.mpg.de), Platform for RIKEN Metabolomics (http://prime.psc.riken .jp), MassBank (http://www.massbank.jp/?lang=en), Biological Magnetic Resonance Data Bank (http://www.bmrb.wisc.edu), and Human Metabolome Database (http://www.hmdb.ca). Additionally, various standards for reporting this type of data have been proposed over the years, which provide useful pointers on the metadata required.[28] These initiatives demonstrate how the development of metadata stimulates debate around the conditions under which data can travel, including the ways in which the circumstances of data production should be described to enable users to compare such information with their existing laboratory experience.

4.3 Distributed Reasoning in Data Journeys

The debates surrounding the codification of embodied knowledge involved in data production may be interpreted as vindicating Polanyi's insistence on its tacit quality. This is knowledge that cannot be acquired without personal involvement in, and extended familiarity with, certain types of material intervention. Learning happens as much through reasoning as it does through doing, and experimental intervention remains the most important form of doing within the biological sciences.[29] Both of the examples I discussed point to difficulties in using metadata to *replace* users' expertise and experience in evaluating components of experimental research. Even in a highly standardized area as model organism research, where building consensus on specific procedures and materials attracts more attention and resources than in other biological fields, database users do not blindly trust the metadata provided by curators and instead use them as starting point for their own investigation and assessment of the methods and materials used by data producers. Rather than replacing users' expertise and experience in the lab, metadata serve as *prompts* for users to use embodied knowledge to critically assess information about what others have done.

I do not mean this to diminish the value of inquiry carried out *in silico*. On the contrary, noting the links between digital and material research practices points to the numerous ways in which the travel of data through online databases can complement experimental work and to the difficulties of using digital tools toward discovery in the absence of skills acquired by interacting with the material world that is under investigation—in ways that are always mediated by social and institutional settings. It is precisely in shaping the social environments in which research takes place that the standard terms, tools, and infrastructures introduced to make data travel have their strongest impact. The employment of metadata is fostering a new level of reflexivity and communication among researchers about what are, or should be, the salient features of experimental work. It also encourages explicit debate around aspects of experimental practice that otherwise tend to remain unquestioned—such as the comparability of experimental procedures and materials across labs, and the reliability of specific instruments and the ways in which various environmental conditions affect their use. Database curators welcome this way of using metadata as a form of active engagement with data travel.[30] I regard critical debate over potentially different forms of embodied knowledge as an important achievement of data curation, particularly of attempts—however misguided and problematic in terms of their accuracy—to capture and express knowledge otherwise kept "personal" and "tacit." Again, it is important to point out that this is not a completely novel feature in biological research, as debates over the adoption of specific experimental materials, instruments, and procedures have a long history. However, the public nature of metadata—their high visibility, their potential for diffusion, and the sheer ambition to make the embodied knowledge involved in experiments accessible to a wide variety of stakeholders—brings attention to these debates and has contributed to a renewed interested in the principle of experimental replicability and the conditions under which it can be achieved.[31] This in turn facilitated scientific communication about embodied knowledge on a scale never before witnessed in the history of science.

Discussions and decisions over how to codify embodied knowledge are particularly useful given the growing number of biologists involved in data journeys and recent shifts in the scale of biological collaborations.[32] I already stressed how data communication and reuse, particularly within model organism communities that include thousands of research groups scattered around the globe, rarely involve personal acquaintance and direct contact between individuals. Furthermore, data journeys increasingly involve a wide variety of experts coming from different fields, some of which— such as computer scientists, statisticians, mathematical modelers, and technicians working in stock centers—do not necessarily have experience in experimental interactions on actual organisms, and yet participate in the chain of processes that make it possible to assign evidential value to the data being disseminated.[33] In such a context, metadata make it possible to think of embodied knowledge as something that is not necessarily restricted to the boundaries of one specific research setting and to the experience of one individual. While a researcher's existing familiarity with experimental techniques, materials, and instruments determines the degree to which she can interpret metadata (and thus assess the evidential value of data), this does not prevent such knowledge from being discussed and articulated. Divergences among the skills and commitments favored within different settings should be identified, and their identification is an important step toward their critical evaluation across research contexts. Thanks to tools like metadata, people with different training and experimental backgrounds can form opinions on each other's practices and use those opinions to interpret data found in silico.

A crucial insight emerging from this analysis is that the embodied knowledge necessary to interpret data does not need to be harbored in the same way by each and every scientist involved in the complex process of producing, disseminating, and interpreting data. Quite the opposite is true: the generative power of data-centric biology partly derives from the opportunity to assess the evidential value of data from a variety of viewpoints, including diverse theoretical backgrounds, experimental traditions, and disciplinary training. This multitude of perspectives affects both the embodied and the propositional knowledge involved in data-intensive processes of discovery. This form of research bears testament to the fruitfulness and creative potential of scientific pluralism, which has been defended by philosophers such as Helen Longino, Sandra Mitchell, Ronald Giere, Werner Callebaut, and Hasok Chang.[34] Because of the significance of embodied knowledge and material practices in my account, my position is most closely aligned with Chang's account of interactive pluralism, which involves "pluri-axial systems of practice" encompassing all elements of research, including data, models, and a variety of possible aims.[35] As Chang remarks with reference to twentieth-century chemistry, "Scientists develop various systems of practice containing different theories, which are suited to the achievement of different epistemic aims. In this way the great achievements of science come from *cultivating* underdetermination, not by getting rid of it."[36] Indeed, one of the motors and attractions of data-centric biology is its potential to include a wide variety of regimes of practice at work in the life sciences, while also helping them confront and challenge each other without simultaneously attempting to homogenize or unify them. Within model organism research, database curators explicitly recognize and exploit scientific pluralism, and many of their packaging procedures are geared to harness the richness and multiplicity of biological research. Exposing data to a variety of epistemic cultures is what makes it possible to expand the evidential value of data beyond that imagined by the original data producers.

This argument sheds light on the ways in which data journeys facilitated through digital technologies are fostering novel forms of collaboration and division of labor in the life sciences. As I stressed in the previous chapter, researchers make important choices at all stages of data handling, including how to format, visualize, mine, and interpret data—and what counts as data in the first place. For any given datasets, several individuals, sometimes hundreds of them, are involved in making those decisions. Thanks to the integrative platforms provided by computers and Internet access, as well as the regulatory and institutional structures I discussed in chapter 2, those individuals are likely to have little in common: they may not know each other and are likely to have different expertises and priorities. Most importantly, each of them might possess a different form of embodied knowledge and thus make use of different skills, methods, and commitments when handling data. In the past, individuals pertaining to such disconnected communities would rarely have crossed paths. Thanks to computational tools, web-based communication, and the increasingly globalized nature of research networks, institutions, and markets, the division of labor within science is becoming more fluid.

As data journey across laboratories all around the world, their life can become so long and unpredictable that there is no way to control who is manipulating them, how, and to which effect. As a result, any one interpretation of data may be achieved through the efforts of a (sometimes very large) group of different individuals with diverse goals—and it is this mix of diversity and cooperation that makes it possible to extract several insights from the same datasets. In such a situation, the intuition of scientific understanding as resulting from the heroic efforts of a lone genius needs to be abandoned. The ability to assign evidential value to biological data is generated not through an overarching synthesis but rather through the fragmented efforts of several different groups of researchers, which offers unique opportunities for integration and cross-pollination (again, a result of de/recontextualization). This reading of data-centric biology as a collective endeavor comes close to Ronald Giere's reading of research at CERN as a large distributed cognitive system.[37] We differ, however, in our emphasis. While Giere is interested in exploring the role played by artifacts in extending human cognition, I wish to stress the distributed nature of understanding itself as a cognitive achievement of scientific collectives.[38] Viewed in this way, distributed reasoning has become a key component of twenty-first-century research, fostering an increasingly pluralistic understanding of data and, as a result, a richer understanding of the natural world.

The presence of vast distributed systems do not undermine the value of individual understanding and synthesis, which still constitutes an important ingredient in data interpretation, not least because of the significant role played by embodied knowledge in this process. Furthermore, individual feedback is crucial to ensure that these systems continue to harness the plurality of perspectives featured in biological research. This is by no means an obvious feature. Indeed, many data dissemination systems may be accused of fostering the opposite tendency: that of hiding, rather than making apparent, the biases, assumptions, and individual motivations introduced at different stages of data journeys, thus transforming these systems into giant black-boxes that need to be accepted at face value, rather than being probed and challenged at every step. This becomes a worrisome and realistic prospect, particularly when we consider that the more complex and distributed data journeys become, the more complex and distributed processes of decontextualization tend to be, with data being manipulated and modified well beyond their format at the moment of generation. Accordingly, it becomes important for database users to understand not only the experimental circumstances in which data were originally produced but also the subsequent steps taken to package data for travel. At least in principle, recontextualizing data for interpretation should involve the ability to reconstruct all the passages through which data have been modified in the course of their travels and thus all the contributions that led to data being visualized and retrieved in a given way—which in turn makes it possible to critically assess the decisions taken at different moments of the journeys, identify eventual errors and misbehaviors, and hold specific individuals or groups accountable. And yet it will by now be clear that, in practice, the complexity and diversity of data dissemination systems are making it impossible for any single individual to reconstruct and understand data travels in such an accurate way.

This brings me to reflect on an important tension characterizing data-centric biology. While data dissemination systems undoubtedly enhance individuals' opportunities to access and evaluate the results of others, they also bring new constraints on the conditions under which critical judgment can be formulated and expressed. As I detail in chapter 6, the distributed nature of data dissemination systems opens up novel opportunities for data integration and interpretation, and yet it limits the ability of any one individual to understand those systems as a whole and requires researchers to place increasing levels of trust in the quality and the reliability of the contributions made by other participants in these networks. I have shown how important it is for database users to exercise their own judgment in assessing data provenance and how curators recognize this need when developing data infrastructures. Nevertheless, both users and curators need to rely on the descriptions of experimental processes that are provided by data producers. Most database curators and users have neither the means nor the time to verify those descriptions, and so the information provided by data producers is typically accepted as correct, unless specific reasons for doubt emerge.[39] In many cases, curators acknowledge the need to take some responsibility in assessing the reliability of data producers. One attempt in this direction is to complement the provision of metadata with curators' own "confidence rankings," which aim to give database users a cursory view of the degree of trust that curators place on the data at hand, so that users can gain a sense of which data are most likely to be reliable before delving into a close examination of their provenance.[40] Reliance on confidence rankings involves delegating an important aspect of the evaluation of the evidential significance data to curators. This is an unavoidable aspect of data journeys and an exemplification of the essentially distributed nature of data-centric reasoning and interpretation in experimental biology. At the same time, this makes it impossible for any one person to provide an overarching assessment of the reliability of these efforts, as any critical judgment on these systems is itself distributed among various kinds of experts who are qualified to judge only certain aspects of the whole enterprise.[41] This situation raises worries as to the long-term sustainability of data infrastructures and of the scientific networks that emerge serendipitously and unpredictably from common engagement with the same datasets.

4.4 Dreams of Automation and Replicability

Given the distributed nature of data journeys, ways to codify and communicate embodied knowledge are urgently needed in order to sensitize all researchers involved to the specific nature of experimental interactions with living matter. The principle underlying the use of metadata is that the characteristics of the embodied knowledge involved in data production can and should be articulated, so that scientists who are not directly involved in that process can still form an impression of how it was carried out. The extent to which such judgments affect data interpretation varies depending on researchers' role in data journeys and the extent to which they share skills and experience in laboratory practices. Those who are mostly involved in data dissemination, such as curators and computer scientists, use metadata to get a sense of how data should be packaged and visualized; the ability to assess other researchers' experimental techniques is arguably less relevant to this kind of work. By contrast, those who aim to reuse data toward new biological discoveries, such as database users, access metadata to get a sense of how data should be valued and interpreted; this is where being able to compare one's embodied knowledge with representations of the embodied knowledge of others becomes crucial to scientific inquiry. This finding illustrates how important the intersections between propositional and embodied knowledge—and how they are socially, materially, and institutionally facilitated and managed—are to gaining a scientific understanding of phenomena.[42]

The idea of carrying out research entirely in silico, without complementing it with experimental intervention on actual organisms, becomes untenable given this insight. This brings me to briefly consider the role of automated reasoning in data-centric biology. The automation of reasoning processes has been a Holy Grail for computational science and artificial intelligence since several decades, and the life sciences constitute an ideal test case to probe the extent to which the limits and costs of human intervention can be overcome through reliance on machines.[43] Following in this tradition, several commentators have promoted the idea that data-intensive efforts may facilitate the complete elimination of human intervention (and thus of manual labor and subjective decision making) from data analysis, resulting in the automatic generation of scientific hypotheses: a fully automated "machine science."[44] This idea goes hand-in-hand with the suggestion that using computer software and statistical analysis to extract meaningful patterns from data will eventually result in the "end of theory," in the sense of making human supervision and conceptualization obsolete to the process of scientific discovery.[45]

What I hope to have shown in this chapter is that new methods of data dissemination and analysis facilitate data journeys not by eliminating human decision making but rather by acknowledging the richness in perspectives and types of knowledge used to form human judgments and using this as a starting point to foster data travel. The ways in which curators mediate and manage the requirements of database users are crucial to biologists' ability to use informational resources to inform experimental work. At the same time, curatorial attempts to standardize and automate data journeys needs to be constantly monitored, challenged, and updated to reflect the latest advances and diversity of approaches characterizing each biological subfield. The full automation of reasoning about data, leading to a data-driven "machine science," remains unlikely despite the impressive advancement in data handling, text mining, and related computational technologies of recent years. The possibility of providing multiple interpretations of the same dataset, so highly valued within contemporary science, is tightly linked to efforts to facilitate the recontextualization of data within different research situations. In turn, this is supported by embodied knowledge that can only partially be stored or reproduced through machines: human agents remain a key component of the sociotechnical system, or regime of practices, that is data-centric biology. Data interpretation is at least partly a matter of understanding the circumstances in which data have been produced—and yet there is no single (or even a "best") interpretation at stake. Depending on their research context and degree of familiarity with specific methods of data production, biologists may interpret the same data in different ways, which fosters the evidential value of data and enhances their chance to contribute to the advancement of scientific knowledge.

This also problematizes the idea that developing reliable metadata helps to foster experimental replicability across biological subfields, as claimed by many curators, experimenters and Open Science activists who are involved in these efforts (as we have seen in the cases of JoVE and BioSharing). I take experimental replicability to denote the ability to reproduce the data generated through a given experiment by following the same procedures.[46] Many scientists value replicability as proof that experimental results are not incidental but rather can be obtained by anybody who has access to a given set of materials, instruments, and procedural knowledge should be able to obtain. Replicability thus acts as an important regulatory goal in science: it demonstrates that data production does not depend on the specific location, personalities, and values of the researchers involved, thus underscoring the reliability, objectivity, and legitimacy of scientific methods. This understanding of experimentation clashes, however, with the persistent finding that the vast majority of experiments are never replicated, and even in the few cases where replication is attempted, it is not always successful.[47] Some Open Science advocates claim that these failures are due to the inaccuracy with which experimenters report on their methods and procedures, which in turn makes it impossible for others to reproduce their laboratory activities. This is why metadata are touted as a remedy: improving the ways in which experimenters capture their actions is expected to increase the replicability of the data thus produced.

In this chapter, I have discussed several reasons why the development of metadata may indeed help improve scientific methods, particularly their potential to facilitate constructive dialogue and feedback on experimental practices across different groups and research traditions. These same arguments lead me to be cautious about the expectation that metadata can enhance experimental replicability, particularly when this is attempted across different biological subfields or research groups using different types of organisms. Depending on their own training and expertise, the ways in which researchers with different backgrounds will understand and act on information provided via metadata may vary widely. While accurate reporting standards do improve researchers' understanding of what others do in their own labs, I have discussed how such understanding is filtered and mediated by their own experiences at the bench, often resulting in diverging assessments of the usefulness and/or significance of specific procedures. This makes it hard to believe that researchers coming from different research traditions will be able to replicate each other's experiments purely on the basis of consulting metadata. What seems more plausible is that researchers working in the same field and on similar organisms may use metadata to improve their likelihood to replicate aspects of each other's work that are of particular significance with respect to a given inquiry. This is an important achievement, which will arguably motivate experimenters to think more carefully about how to report their methods and choose metadata to describe the circumstances of their work, but it is hardly the watershed envisaged by the most optimistic defenders of this idea.

5

What Counts as Theory?

Data cannot be stored and circulated without employing organizing principles. We have seen how this basic requirement is even more pressing when posting data online. Data stored in digital databases need to be standardized, ordered, and visualized, so that scientists can retrieve them in ways that inform their own research. As noted by Maureen O'Malley and Orkun Soyer, the integration of data into a single body of information involves imposing some degree of homogeneity and order over the data in question.[1] At the same time, the drive toward unification and synthesis can overdetermine the ways in which data may be retrieved and interpreted, and indeed in the previous chapter, I showed how database curators attempt to document, rather than conceal, discrepancies in the quality and provenance of data. In this chapter, I thus turn to the organizing principles required to assemble and retrieve data in a database, particularly to the labels used by curators to classify data for the purposes of dissemination. After describing this process of classification in more detail, I argue that it illustrates how data curation activities contribute to the production of knowledge in biology, thus countering the characterization of this work as mere stamp collecting or technical "service" to the biological community. I characterize the activity of labeling data for travel as *theory making*; and I posit that its products, such as bio-ontologies, developmental stages, or taxonomic systems, constitute *clas-*

sificatory theories that can be clearly distinguished from other forms of theorizing typically discussed within the philosophy of science. Finally, I discuss the ways in which my notion of classificatory theory informs broader discussions of the role of theory in data-centric science, particularly of the distinction, often cited in scientific circles, between data-driven and hypothesis-driven research. Paying attention to the theory-making qualities of data curation is a way to emphasize its integral role in processes of discovery, where conceptual and practical decisions about how to visualize data affect the form and quality of knowledge obtained as a result. It is also a way to specify how theory enters data-centric modes of research, without however driving them in the same way as when data are produced in order to test a given hypothesis. This clarifies how data-centric research differs from a supposedly theory-free inductive exercise, while at the same time distinguishing the use of theory in this approach from that characterizing other forms of scientific inquiry.

Before proceeding, I should note that my discussion is grounded on a pluralistic and pragmatic understanding of scientific theorizing, which investigates the function of theory within specific research practices and for particular purposes.[2] I do not believe that philosophical debates should be tied to a uniform definition of theory that holds for the whole of science, and I thus do not see the recognition of variability among types of theories, each structured and used to serve specific research goals, as problematic. Paradoxically, the recent insistence on the plurality and scope of different forms of modeling has threatened to discard the notion of theory altogether as a relevant component of scientific practice, particularly in the case of philosophers sympathetic to the semantic view of theories as families of models[3] or models as autonomous agents.[4] By contrast, I am convinced that theory has an important role to play in scientific epistemology and yet that our understanding of what counts as theory should reflect the turn to scientific practice characterizing philosophical research on models.

5.1 Classifying Data for Travel

In chapters 1, 3, and 4, we have seen how the fast development of data infrastructure has brought new urgency to the need for common labels under which data gathered across a variety of contexts can be classified and retrieved. We have also seen how challenging this requirement is, given that biology is a highly fragmented field, encompassing numerous epistemic cultures with diverse commitments, interests, and research methods. Differences among epistemic cultures may shift rapidly depending on the alliances developed to study a specific topic, as epistemic cultures form or dissolve on the basis of which projects are funded and which collaborations are regarded as fruitful in the long term.[5] Despite this volatility, differences among epistemic cultures often manifest themselves through the elaboration of local terminologies to refer to biological objects and practices. For instance, what ecologists see as a symbiont might be classified as a parasite by immunologists; and molecular and evolutionary biologists tend to attribute different meanings to the term "gene."[6]

The use of language is unavoidable when trying to classify large masses of data for dissemination, and yet making data travel across communities that use terms in different ways is no small feat. As highlighted by several Science and Technology Studies (STS) scholars, any classification system has a stabilizing force, and such stability is needed to search and retrieve data from databases.[7] However, this requirement clashes with the instability characterizing the user community, as observed in a highly cited review in bioinformatics: "This is one of the most important general problems in building standards for biology—our understanding of living systems is constantly developing."[8] A classification fit to support data journeys needs to be *dynamic* enough to support the ever-changing understanding of nature acquired by the biologists who use it and at the same time *stable* enough to enable data of various sorts and significance to be quickly surveyed and retrieved. Can data classification through standard categories enable data-centric research without at the same time stifling its development and pluralism?

In the late 1990s, biologist Michael Ashburner proposed an answer to this question in the form of a functional approach to data classification. Ashburner's idea, first presented at the Montreal International Conference on Intelligent Systems for Molecular Biology in July 1998, was to classify data on the basis of the biological entities and processes that genomic data were used to research. This view was born out of Ashburner's experience in developing one of the first community databases (FlyBase, for the fruit fly *Drosophila melanogaster*) and was shared by many other database developers interested in serving "not just organism-specific communities, but also pharmaceutical industries, human geneticists, and biologists interested in many organisms, not just one."[9] It involved classifying data gathered on each gene according to the known molecular function and biological role of that gene. The implication was that the terms used for data classification should be the ones used by biologists to describe their research interests— that is, terms referring to biological phenomena. Thus, for instance, a database user wishing to investigate cell metabolism should be able to type "cell metabolism" into a search engine and retrieve all available genomic data of relevance to her research.[10]

This approach was implemented as an "ontology," following a strategy for ordering and storing information that was already popular in computer science and information technology, because it enables programmers to produce a formal representation of a set of concepts and of the relationships among those concepts within a given domain. I should specify from the outset that the choice of the word ontology in this context has little to do with the long tradition in the philosophical study of being.[11] Rather, it has to do with signaling to other scientists and to the world that what is at stake in the development of labels is the very core of scientific and technological innovation—that is, the map of reality used by scientists to coordinate efforts and share resources. This map needs to be drawn on the basis of pragmatic considerations, rather than theory or ideology. As remarked in a review of standardization efforts in model organism biology, "there is a considerable difference between building a "perfect ontology" for knowledge representation, and building a practical standard that can be taken up by the entire community as a means for information exchange. If the ontology is complex, it is unlikely that the wider community will use it consistently, if they use it at all."[12] When applied in the biological domain, each label is used to refer to an actual biological entity and, at the same time, to classify available data. Thus were born the bio-ontologies, defined as "formal representations of areas of knowledge . . . that can be linked to molecular databases" and thus can be used as classification systems for data sharing and retrieval.[13]

The first bio-ontology to achieve prominence among databases was the Gene Ontology, or GO, whose general features I discussed in chapters 1 and 2. Here I wish to focus on its classificatory role as crucial to uncovering the epistemic structure and modus operandi of data-intensive science, particularly the role of theory within it. GO was developed as a standard for the classification of gene products. It encompasses three different ontologies, each mapping a different set of phenomena: a *process* ontology describing "biological objectives to which the gene or gene product contributes,"[14] such as metabolism or signal transduction; a *molecular function* ontology representing the biochemical activities of gene products, such as the biological functions of specific proteins; and a *cellular component* ontology, referring to the places in the cell where a gene product is active (nuclear membrane or ribosome).[15] GO terms are related through a network structure. The basic relationship between terms is called containment and it involves a *parent* term and a *child* term. The child term is *contained* by the parent term when the child term represents a more specific category of the parent term. This relationship is fundamental to the organization of the bio-ontology network, as it supports a hierarchical ordering of the

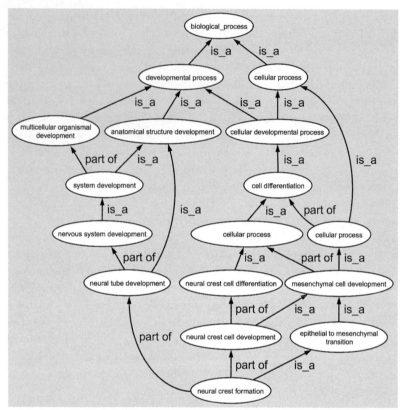

FIGURE 6 Visualization of one part of the Gene Ontology as a network of terms. (Gene Ontology website, accessed April 2010 and slightly modified by Michel Durinx to emphasize the relationships among terms, http://geneontology.org/page/download-ontology#go.obo_and_go.owl.)

terms used. The criteria used to order terms are chosen in relation to the characteristics of the phenomena captured within each bio-ontology. For example, GO uses three types of relations among terms: "is_a," "part_of," and "—regulates."[16] The first category denotes relations of identity, as in "the nuclear membrane is a membrane"; the second category denotes mereological relations, such as "the membrane is part of the cell"; and the third category signals regulatory roles, as in "induction of apoptosis regulates apoptosis." In other bio-ontologies, the categories of relations available can be more numerous and complex: for instance, including relations signaling measurement ("measured_as") or belonging ("of_a"). One way to visualize bio-ontologies is to focus on their hierarchical structure as a network of terms, as illustrated in figure 6. This approach is the one preferred by experimental scientists, who use it to focus on the ways in which terms are related to each other.

GO is now incorporated into most community databases for model organisms, including WormBase, the Zebrafish Information Network, DictiBase, the Rat Genome Database, FlyBase, The Arabidopsis Information Resource, Gramene, and the Mouse Genome Database.[17] Many other ontologies have appeared since its creation. Some are devoted to classifying data gathered on a given type of object, such as the Cell Ontology or the Plant Ontology; others focus on data gathered through specific practices, such as the Ontology for Clinical Investigation and the Ontology for Biomedical Investigation.[18] I am restricting my examination to the bio-ontologies collected by the Open Biomedical Ontologies (OBO) consortium, an organization founded to facilitate communication and coherence among bio-ontologies with broadly similar characteristics.[19] GO is particularly relevant to my argument not only because of its foundational status among bio-ontologies but also because, as I specified in chapter 1, it has been explicitly developed to facilitate data integration, comparison and reuse across model organism communities. While bio-ontologies may be used for other endeavors, such as for instance managing and accessing data even before they are circulated beyond the laboratory where they are originally produced,[20] GO is specifically devoted to expressing the biological knowledge underlying the reuse of data within new research contexts: it captures the assumptions that researchers need to share to successfully draw new inferences from existing datasets.[21] This becomes clear when considering some of the processes through which bio-ontologies are created, such as the following:

- The *selection* of terms to be used as keywords for data searches: Terms are chosen for their popularity within current scientific literature and thus their intelligibility to potential users of databases. As emphasized by the GO founders, "One of the factors that account for GO's success is that it originated from within the biological community rather than being created and subsequently imposed by external knowledge engineers."[22] Curators are encouraged to pick terms in use within current scientific literature, so as to make sure that they do not insert their own pet terms in the classification system. Further, they are responsible for compiling lists of synonyms for each term, so that research communities using different terms for the same entity would still be able to access the wished-for data.
- The *definition* of what the terms actually mean: Bio-ontologies are not simply lists of keywords. Rather,

they are "controlled vocabularies": collections of terms whose definitions and relations to each other are clearly outlined according to specific rules. The meaning of each term is unambiguously specified via a definition in which curators specify the characteristics of the phenomenon that the term is intended to designate.[23] For instance, the GO defines the term "ADP metabolism" as "the chemical reactions and pathways involving ADP, adenosine 5'-diphosphate."[24] The definition of terms matters greatly to the success of bio-ontologies as classification systems. Researchers can only make sense of data retrieved through a bio-ontology label if they know precisely which entity that label refers to.

- The *mapping* of terms to specific datasets: Datasets do not come with a ready-made tag indicating the research areas in which they might prove relevant—and thus the bio-ontology terms that will be used for their classification. The mapping of data to bio-ontology terms needs to be done manually, through a process called "annotation," either by curators or by data producers themselves. Again, how curators decide to label data has a strong impact on the functioning of bio-ontologies for data retrieval. Database users need to trust that the data classified under a specific term are actually relevant as evidence for the investigation of the related phenomenon.

These three processes are subject to regular revision, with curators well aware that the knowledge captured by bio-ontologies is bound to change with further research. The advantage of bio-ontologies as digital tools is that they can be updated to reflect developments in the relevant scientific fields.[25] They are built to be a dynamic rather than a static classification system: "By coordinating the development of the ontology with the creation of annotations rooted in the experimental literature, the validity of the types and relationships in the ontology is continually checked against the real-world instances observed in experiments."[26]

A philosophically interesting way to view bio-ontologies is to conceive of them as a series of descriptive propositions about biological entities and processes. So in the example provided in figure 6, the bio-ontology tells us that "cell development is part of cell differentiation" and "cell development is a cellular process." One can assign meaning to these statements by appealing to the definitions given to the terms *cell development, cell differentiation*, and *cellular process*, as well as to the relations "is_a" and "part_of." Similarly, one learns that "neural crest formation is part of neural crest cell development" and that "neural crest formation is an epithelial to mesenchymal transition." Viewed in this way, bio-ontologies consist of a series of claims about phenomena: a text outlining what is known about the entities and processes targeted by the bio-ontology. As in the case of any other text, the interpretation given to these claims depends partly on the interpretation given to the definitions assigned to the terms and relations used. The hierarchical structure given to the terms implies that changes to the definition of one term, or to the relationship linking two terms in the network, have the potential to shift the meaning of all other claims in the same network.

Yet definitions are not the only tool available to researchers to interpret claims contained in a bio-ontology. Claims are also assessed through the evaluation of the quality and reliability of the data that has been linked to each of the terms involved. As detailed in the previous chapter, this is possible through the consultation of metadata, which put users in a position to evaluate the interpretation given to the data by their producers (if any) and the way in which they support (or not) the claims made in the bio-ontology. For example, a researcher interested in neural crest formation can investigate the validity of the claim "neural crest formation is an epithelial to mesenchymal transition" by checking which data were used to validate that assertion (i.e., to link the term "neural crest formation" with the term "epithelial to mesenchymal transition" via the relation "is_a"), what organism they were extracted from, which procedures were used to obtain them, and who conducted that research. As a result, researchers can appeal to their own understanding of what constitutes a reliable experimental setup, a trustworthy research group, and an adequate model organism to interpret the quality of the data in question as evidence for a claim.

5.2 Bio-Ontologies as Classificatory Theories

Much philosophical work on the characteristics and role of scientific theories is grounded on the intuition that constructing a theory implies introducing a new language to talk about the natural world. This idea resonates with the Kuhnian understanding of incommensurability among paradigms, according to which paradigm shifts involve language shifts and even when key concepts stay the same, their meaning tends to differ.[27] In this section, I argue that bio-ontologies represent a counterexample to this intuition. They constitute a form of scientific theorizing that has the potential to affect the direction and practice of experimental biology in the long term, and yet they do not aim to introduce new language into biological discourse but rather attempt to gather and express consensus on what constitutes established knowledge. In other words, bio-ontologies are an instance of how theory can emerge from the attempt to classify entities in the world, rather than from the attempt to explain phenomena. Recognizing this role for bio-ontologies involves accepting that descriptive predicates can all have theoretical value, regardless of the generality and the novelty of the concepts that they contain—regardless, that is, of whether they can be clearly differentiated from empirical observations and whether they challenge existing knowledge claims. As I intend to show, whether descriptive predicates constitute a form of theory depends on the role that they play within the life cycle of experimental research. Furthering the arguments presented in the previous chapter, I stress that recognizing this role for bio-ontologies sheds light on the interactions between theory making and experimentation that characterize data-centric research. Viewing bio-ontologies as theories illustrates how the theoretical claims underlying data classification can only be interpreted through consultation of the evidence supporting those claims and thus from familiarity with the experimental practices through which that evidence has been produced.

As a starting point, I use Mary Hesse's network model of scientific theories, within which theories are defined as networks of interrelated terms. The meaning of each term depends both on the definition given to the phenomenon to which the term applies and on its relation to other terms. Relations among terms are expressed by way of law-like statements (for instance, Newtonian mechanics includes the terms *mass* and *force*, which are related by the law-like statement "force is proportional to mass times acceleration"). Theories can thus be propositionally expressed by enunciating the series of law-like statements defining the relations among the terms used to refer to phenomena. Within this framework, Hesse argues that "there is no distinction in kind between a theoretical and an observation language."[28] What Hesse means is that scientific language is replete with descriptive statements, whose truth depends on two fundamental factors: (1) their relation to other statements used in science to describe related phenomena (what Hesse calls "the coherence of the network of theory'"), and (2) their relation to various kinds of evidence (in her words, "empirical input").[29] These descriptive statements may vary greatly with respect to the scope of phenomena to which they may be applied, the degree of generality of the concepts that they incorporate, and their degree of abstraction from specific empirical instances. Nevertheless, there is no obvious way to divide those statements in two distinct epistemic classes, one "theoretical" and one "observational." The consequence of this realization is that all descriptive statements have a theoretical value and can be treated as theory depending on their context of use.

This picture of how theory works in experimental science neatly fits the case of bio-ontologies, which are networks of terms defined by a predicate (the definition of the term) as well as by a specified set of empirical evidence (the data classified under that term). The terms are related so as to form descriptive sentences applicable to a variety of empirical situations, such as "a nucleus is part of a cell." The truth of such statements depends on the truth of the definitions provided for each of the terms used, on the relation stipulated between each statement and the rest of the network of which it is part, and on its relation with available evidence as gauged by scientists. Descriptive sentences captured in bio-ontologies have the same epistemic role as testable hypotheses: they are theoretical statements whose validity and meaning can only be interpreted and assessed with reference to empirical evidence and the conditions under which that evidence has been produced. Some of the biologists with whom I discussed this view initially resisted my characterization of bio-ontologies as forms of theory. They understood biological theories as needing to express something that is not yet empirically established or that is very new, rather than what is already known. At the same time, they agreed that bio-ontologies are supposed to include "all we know" about biology and thus fit a view of theory as expressing existing knowledge about biological phenomena (including, but not limited to, newly available knowledge). The following passage, taken from an e-mail sent to me by the director of a major center for the elaboration of bio-ontologies, illustrates the point: "The relationships we capture tend to be the more basic and universally accepted ones, not those that are currently evolving as biological knowledge is extended. Perhaps that is why biologists do not think of database design as the highest level of our field, analogous to a mathematical model for the universe. I think ontology structure is closer to a true conceptual model, but even there the relationships captured are not on the cutting edge of biology but those that are already generally agreed on and not controversial."[30]

Hesse argues that the choice, use, and modification of each term and relation expressed in a network depends on the empirical study of the phenomena that those terms and relations are supposed to describe and thus on the complex processes of experimentation, intervention, and classification involved in empirical research.[31] Thus the theoretical representation of knowledge is instrumental to the goals and needs of empirical research. Similarly, the main motivation behind the development of bio-ontologies is to facilitate data integration to foster further research. The nature of the data that are available for inclusion, as well as their format and the methods used by the laboratories that collected them, inform all decisions about which terms to use, how to define them, and how to relate them to each other. In line with Hesse's arguments, terms are not defined on the basis of purely theoretical considerations, but they are formulated to fit the studies actually carried out by empirical scientists. Terms referring to entities and processes that are not under investigation, which therefore do not have data associated with them, are not incorporated in bio-ontologies. Similarly, definitions are often formulated through the observational language used by empirical researchers, rather than on the basis of their coherence with each other or any one given framework.

This became clear to me when attending a GO Content Meeting dedicated to the terms *metabolism* and *pathogen*, two notions that are notoriously difficult to define and are treated differently depending on the subdiscipline of interest.[32] Discussion among the immunologists, geneticists, and developmental and molecular biologists attending the meeting started from a tentative definition of the terms and progressed by correcting that initial definition so that it would fit counterexamples. Most counterexamples were provided *in the form of actual observations* from the bench or the field. For instance, the proposal that pathogens be treated as an independent category from organelles, which was popular among immunologists, was dismissed by ecologists and physiologists on the basis of cases of endosymbiosis where pathogens turn out to be both symbionts and parasites of the same organism. In these cases, it was argued that pathogens cannot be treated as a separate, independent category from other microscopic components of the host's cell, since they also play a role contributing toward the well-functioning of the cell as a whole. According to specialists in endosymbiosis and its role toward plant development, these pathogens should therefore figure as "part_of" the cell, rather than as a separate entity with no relations to it. Certain kinds of bacteria (such as nitrogen-fixing bacteria) can have at the same time mutualistic and parasitic associations with their hosts.[33] Thus on the basis of observed cases and heated discussions among database users and curators, a whole theoretical category (the one of "pathogen") was modified to fit a different context and definition.

This example shows how one of the keys to the successful functioning of bio-ontologies is the avoidance of distinctions between theoretical and observational language and related concerns. Bio-ontologies are meant to represent knowledge about existing phenomena as currently investigated, rather than to make sense of them through detailed explanations or daring new hypotheses. For that purpose, it does not matter that bio-ontologies do not, and were never intended to, introduce new theoretical terms to explain biological phenomena. Not all scientific theories operate in that way, nor, as

I discuss below, are they all meant as sweeping perspectives unifying a whole discipline. Bio-ontologies are better understood as what I call *classificatory theories*. They emerge from a classificatory effort, in the sense that they aim to represent the body of knowledge available in a given field to enable the dissemination and retrieval of research materials within it; are subject to systematic scrutiny and interpretation on the basis of empirical evidence; affect the ways in which research in that field is discussed and conducted in the long term; and, most importantly, express the conceptual significance of the results gathered through empirical research. Researchers who use bio-ontologies for data retrieval implicitly accept, even if they might not be aware of it, the definition of biological entities and processes contained within bio-ontologies at the moment at which they are consulted, which in turn affects how data are used in subsequent research. It is because of this crucial role in expressing available knowledge of biological phenomena, and thus guiding and structuring subsequent research, that bio-ontologies are best regarded as a form of theory.

This view of theory is not a newcomer within the philosophy of science. It parallels Claude Bernard's idea of theory as a "stairway" toward the establishment of new scientific facts: "By climbing, science widens its horizons more and more, because theories embody and necessarily include proportionately more facts as they advance."[34] Theoretical claims are what bring us from one observation to the next and make it possible to develop complex explanations; descriptive statements such as those used to classify biological organisms within the taxonomic tradition can thus be seen as theoretical.[35] This claim builds on previous claims about the profound conceptual implications of the choosing taxonomies, especially concerning the conceptualization of organisms in biology, as put forward chiefly by John Dupré.[36] He argued that the classification of species and biological individuals has "a life of its own," with important consequences for biological theory,[37] and that classification systems such as taxonomies and phylogenies profoundly affect the ways in which organisms are conceptualized.[38] Here I go one step further, by claiming that such classifications may sometimes *constitute* theory in biology.

It is important to note that I am not arguing for all scientific classifications to be regarded as theories. Rather, I am pointing out that some classifications can play the role of theories depending on the extent to which they embody and express a specific interpretation of the overall significance of a set of empirical results. Whether classifications other than bio-ontologies, such as the ones used in the natural history tradition or contemporary taxonomy, fit this requirement remains a topic for further debate. Staffan Müller-Wille's reading of the classificatory practices used by Linnaeus, which he views as guided by the need to order and summarize data pouring in Linnaeus's study from all corners of the globe, and yet resulting in a specific interpretation ("ontological scaffold") of the biological significance of those data, could be interpreted as indicating that the Linnaean classification system functioned as a theory in eighteenth-century botany.[39] Similarly, the arguments put forward by Bruno Strasser on the intertwining of experimental and classificatory practices in the construction of databases such as GenBank point to a potentially theoretical role of the classification of protein sequences within 1970s–80s molecular biology.[40]

A comparable case that I wish to consider in more detail is what Alan Love describes as "typological thinking" in developmental biology and systematics, which in his words involves "representing and categorizing natural phenomena, including both grouping and distinguishing these phenomena according to different characteristics, as well as ignoring particular kinds of variation."[41] A good example of this is the choice and use of descriptions and/or representations of normal stages of development as classification tools for data in developmental biology. Love makes a strong argument, built in relation to recent work on the importance of natural kinds as classification tools,[42] that the typologies used to classify stages of development are chosen partly on pragmatic grounds, such as familiarity with the types and formats of relevant data and the strategies through which those data are collected, and partly on the basis of expectations about which aspects of the organisms in question will be of most relevance to advancing existing scientific knowledge about developmental processes. For instance, developmental biologists select images or descriptions exemplifying stages in the postembryonic development of chick or frog fetuses on the basis of their experience of what constitutes a "typical stage" across multitudes of different specimens (which in turn involves familiarity with the ways in which specimens are studied and the specific kinds of data acquired through those processes) and of their background expectations about how images of fetuses at different stages are collected and interpreted in their field.[43]

These are choices about how to acquire data enabling the investigation of biological processes of interest (in this case, the development of a given organism). Notoriously, these choices have both descriptive and normative undertones: "[The classification of] normal stages involve assessments of 'typicality' because of enormous variation in the absolute chronology of different developmental processes."[44] They embody assumptions about how to provide stable and general representations of an essentially dynamic and species-specific process. This results in knowledge about the very processes being investigated (e.g., statements identifying causal connections between developmental stages and homologies between different species), which is in turn used to evaluate and interpret available data. Classificatory labels such as developmental stages embody and formalize these choices and assumptions and thus come to express substantive decisions about how development will be measured; which parameters count, and why; and which terminology and methods best encapsulate available knowledge on organismal development, while at the same time facilitating future efforts of data collection and interpretation. So on the one hand, these classificatory systems work at least partly because of their efficiency as tools to coordinate the collection, dissemination, and analysis of data on development; on the other hand, they constitute themselves conceptualizations of development that have consequences for how research in this area is carried out and on what. As in the case of bio-ontologies, to adequately assess the import and significance of such conceptualizations, biologists typically need to evaluate the experimental context in which they were originally proposed. This provides a way to understand the pragmatic motivations for specific conceptualizations and to assess their usefulness when applied in other experimental contexts.

5.3 The Epistemic Role of Classification

My arguments can be framed as part of a wider, ongoing effort to reconsider the epistemic role of classificatory practices in the sciences and overcome the long-held perception of classification as a conceptually uninteresting part of scientific research. To this aim, Ursula Klein and Wolfgang Lefèvre examined how historical actors used the diversity of classification systems populating eighteenth-century chemistry to highlight wide differences in ontological beliefs.[45] In a similar vein, Lorraine Daston's extensive historical studies of taxonomic practices led her to frame these activities as "metaphysics in action";[46] Albert Crombie and Ian Hacking argued that scientific classification involves a specific style of reasoning that is central to scientific research;[47] and John Pickstone singled out natural history, and the classificatory practices therein, as a unique "way of knowing"—in his words, "knowing the variety of the world" by "collecting, describing and displaying."[48] Pickstone also emphasized the broad historical and cultural context of classification in natural history, which includes a variety of motives such as "pride of possession, intellectual satisfaction and commerce and industry," and stressed that the presence of nonepistemic reasons to classify does not reduce the epistemic value of such activities—just like the material, conceptual, and institutional conditions for data journeys shape their evidential value.

Classification practices such as the ones depicted above coordinate and underlie the tracking of biological phenomena and the collection of the resulting data into representations that can be used to model those phenomena. In doing so, they offer a substantive formalization of biological knowledge about organisms, whose primary goal is to facilitate the interpretation and further collection of data about organismal structures and development across species.[49] These formalizations uncover knowledge that is given for granted when collecting, disseminating, and using data. Furthermore, by assembling, integrating, and expressing knowledge used to handle data, these classification systems make a unique contribution to scientific research. They express knowledge that cannot be found anywhere else (models, instruments, existing theories, or textbooks). This is not necessarily the same as contributing entirely new knowledge in the sense of capturing new discoveries. We have seen how bio-ontologies integrate and formalize knowledge that already exists but is dispersed in different areas of biological research (such as different model organism communities or different subdisciplines working on the same phenomenon). This formalization requires extensive conceptual work and often constitutes an important step forward in the development of biological knowledge, even if it does not necessarily incorporate paradigm-shifting discoveries. For instance, GO has developed a new definition of the term "gametogenesis" that takes account of the differences between this process in plants and animals, which researchers were able to articulate clearly thanks to this classificatory practice.[50]

This analysis demonstrates that classification systems such as GO are not the same as lists, whose main function is to group a given set of items and arrange them in a specific order. The process of ordering alone does not amount to theorizing, particularly in cases where this is done according to preestablished criteria (e.g., in the case of biodiversity surveys, where an inventory is constructed around a specific conceptualization of species, or botany manuals, where classificatory categories aim to enable readers to identify plants in the wild). When ordering is intertwined with the analysis and interpretation of the scientific meaning of the items being sorted, something more is at stake. Within GO, the criteria for what count as good terms, definitions, and links with data, and thus for what biological knowledge is expressed in the system, are developed as part and parcel of the process of classification. As a result, classification systems such as bio-ontologies and developmental stages are more than just theory-laden. They express criteria to evaluate and interpret the scientific significance of the items that they are used to classify. This is knowledge that underpins and directs scientific practice, the production of new data, and, most importantly, debates about the structure and functioning of biological entities and processes.

Viewing these systems as theories recognizes their crucial epistemic role in expressing the biological knowledge underlying experimental research, while at the same time emphasizing the fallible, dynamic, and context-dependent nature of such knowledge. Like Quine's webs of beliefs, classificatory theories face the "tribunal of experience": they are regularly challenged by new evidence and can be modified and updated when necessary. This insight demonstrates how, whether or not philosophers agree on the existence of this form of theory, the identification of classificatory theories can prove valuable to scientists. Indeed, recognizing the role of bio-ontologies as theories uncovers and highlights the conceptual substance and commitments underlying their adoption, thus alerting biologists against an uncritical use of databases that use these classification systems. Similarly, the identification of types used in developmental biology as a form of theory enables their critical discussion and, where necessary, questioning in the face of contradictory conceptualizations or experimental results. As stated by three prominent developmental biologists discussing the status of developmental stages as tools for data analysis in vertebrates, "Thinking in terms of types, either as developmental stages or as putative ancestors, can be helpful in searching for order in the diversity of animal life. However we need to be aware of the limitations of typologism."[51] Recognizing the theoretical contributions embedded in classificatory efforts is one way to acknowledge such limitations and encourage their critical scrutiny.

One important objection might be raised at this point: Why should we regard bio-ontologies as theories, rather than as yet another important component of scientific practice that affects, but does not yet constitute, theoretical knowledge? What is at stake in the claim that such classificatory practices constitute theorizing? This objection seems especially strong in light of recent literature on the role played by elements other than theory, such as models, experiments, and instruments, in shaping scientific research. Why are classifications of the type I described not a hitherto understudied type of model, or even background knowledge, rather than a form of theory? Would it not be enough to point to the important role played by classificatory activities in providing conceptual scaffolding for the evolution of biological knowledge (as implied, for instance, by Wimsatt and Griesemer)?[52]

In response to these questions, what needs to be noted is that bio-ontologies are not simply influencing knowledge-making practices in biology in the way that a model or instrument would. The reason for their epistemic power is that they express, rather than simply affect, the knowledge obtained through scientific research; and they do it in a way that is unique, since such knowledge is not formalized anywhere else in the same way and enables experimenters to make sense of the results they obtain. They therefore play a significant role in delimiting the content and development of knowledge (what counts as biological insight, which form it can take) and as a target for critique and reference points for the construction of alternative accounts. Of course, classificatory theories are best understood in relation to the collection of models, instruments, and commitments made by the researchers who produced them, as in the case of any scientific theory.[53] However, they cannot be reduced to any of those other elements; moreover, they provide a way to link and *evaluate* the epistemic results of using all those methods and tools to research nature. Articulating knowledge that enables scientists to assess and value their results is an achievement that goes well beyond listing a set of commonly used assumptions as a basis for further inquiry. In the latter case, existing knowledge is applied to put a given set of items into some order; in the former, existing knowledge is transformed and developed so as to facilitate the conceptual analysis of data. Conceived in this way, classificatory theory parallels Krakauer et al.'s discussion of what they call "bottom-theory," emerging from data handling practices: "Theory provides the basis for the general synthesis of models, and a means of supporting model comparisons and ideally establishing model equivalence."[54] This is why the results of some classification systems should be viewed as theories rather than mere background knowledge, even if this notion of theory differs from traditional depictions as a series of axioms or principles with great explanatory power and universal scope, as I show in the next section.

5.4 Features of Classificatory Theories

I now consider in more detail what distinguishes classificatory theories from other forms of scientific theorizing, such as law-like generalizations (like the Hardy-Weinberg law of population equilibrium), explanatory principles (such as selection in evolutionary theory), mechanisms (e.g., descriptions of how DNA replication takes place), or modeling activities (for instance, the use of graphs to illustrate metabolic networks, which could count as forms of theorizing under a semantic account). I discuss four characteristics typically attributed to theories in the philosophy of science: the ability to generalize, unify, explain, and provide grand visions guiding future research. I argue that these features are indeed exhibited by classificatory theories but in ways that differ substantially from established accounts.

5.4.1 Generalizing. Classificatory theories aim toward generality in the sense that they provide common labels covering a large number of phenomena and related research results. These labels need to be general in order to be used to interpret the evidential value of new datasets, provide heuristic guidance and conceptual structure to future investigations, and contribute to biologists' understanding of phenomena.[55] Their level of generality is, however, not fixed and disconnected from aspirations to universality:

1. It is *not fixed* because the scope of application of classificatory theories can vary greatly depending on the research context and objects on which they are used, as well as the scientist(s) using them. On the one hand, the significance and epistemic value of classificatory theories is domain dependent: they can be accepted or challenged depending on the research context in which they are used. On the other hand, the meaning and intelligibility of these theories depends on their user's knowledge of the scientific practices through which they were developed—that is, knowledge of the materials, settings, and techniques through which their objects (i.e., the phenomena that they posit and characterize) have been tracked. In other words, the meaning and epistemic value assigned to theoretical claims made as a result of classificatory practices depends on one's expertise in tracking the phenomena in question.[56]

2. The generality of classificatory theories is also *disconnected from aspiration to universality*, because classificatory theories tend to accrue generality over narrowly defined domains. For instance, both GO and stages of development aim to generalize over species, by making classificatory categories applicable beyond the species on which data were originally obtained. This aspiration manifests itself differently depending on which species are targeted. Different classifications of stages of development tend to cover species that are phylogenetically close, such as specific families of vertebrates. Within GO, generalizing over species has meant generalizing over the most popular model organisms on which molecular data are being gathered, so the principal aim is generalization over *Arabidopsis thaliana, Saccharomyces cerevisiae, C. elegans, Drosophila melanogaster,* and *Mus musculus* (no small endeavor of course, given the diversity among these organisms[57]).

The lack of universality and context dependence of classificatory theories raise an important philosophical question—namely, whether claims expressed by this type of theory amount to biological laws and in which sense of lawfulness if at all. To answer this question, I rely on Sandra Mitchell's functional reading of lawfulness in scientific theories, according to which "exception ridden, nonuniversal true generalisations can under clearly defined conditions, function in the same way that universal, exceptionless generalisations do in explanation and prediction."[58] Mitchell adds that in many instances such nonuniversal generalizations are "not as easy to use" as the exceptionless, universal generalization often taken by philosophers to constitute the ideal scientific theory. This observation has often been used by philosophers as evidence for the superior status of universal laws over localized generalization, and yet to my knowledge, no philosopher has yet conducted a survey to verify whether scientists prefer universal to nonuniversal generalizations and under which circumstances.[59] On the contrary, studies such as Mitchell's on experimental biology and Nancy Cartwright's on physics,[60] among many others, demonstrate how localized generalizations may well prove more useful to a working scientist than a set of universal generalizations that are however too abstract to be applied to the phenomena at hand. Building on these insights, I view claims made in classificatory theories such as bio-ontologies as akin to what Mitchell calls "pragmatic laws": a "more inclusive class of regularities" that are typically regarded by biologists as key explanatory resources and contributions to scientific knowledge.

5.4.2 Unifying. Another property often ascribed to theories in the philosophical literature is that of unifying claims about phenomena, thus providing an overarching account of how the world works—the classic model for this being the general theory of relativity, which provides a unified description of gravity as a geometric property of space and time. I wish to argue that classificatory theories have indeed unifying power and yet display it in ways that differ considerably from grand theories such as the general theory of relativity. To clarify this claim, I will make use of a distinction introduced by Margaret Morrison in her 2007 study of theoretical unification. This is the distinction between *synthetic* and *reductive* ideals of unity, where the former captures attempts to provide a single account of the subject matter and the latter aims to establish some kind of commonality between different phenomena, without however necessarily embedding that commonality within an overarching conceptual structure. Systems such as bio-ontologies aim to achieve classificatory categories that highlight similarities between different species, and accordingly bio-ontology curators devote much of their time to standardizing terminology in ways that fit different biological subcultures. This constitutes a nice example of a reductive process of unification, and it is not typically coupled with the attempt to achieve synthetic unity, for instance by providing a single account of what constitutes an organism, under which all GO definitions could fit. Rather, classificatory systems such as bio-ontologies or developmental stages aim to enable scientists to pursue disunified, fragmented research about a large variety of objects. Classification systems tend to achieve reductive unity by abstracting from specific instances of entities or processes to one label pointing to common features. Once this is achieved, classificatory theories strive to capture both the complexity and the diversity of biological phenomena by using a vast number of labels to identify them. GO, for instance, keeps diversifying and adding terms, rather than trying to reduce the terms it uses, so as to capture as precisely as possible existing knowledge about the specific processes or entities being investigated. This kind of classification does not aim to reduce the number of labels that are used to identify phenomena. This might happen as a result of the attempt to develop efficient tools for experimental research or, as we saw above, generalizations over species, but it is not a primary concern nor a necessary characteristic for these systems.

5.4.3 Explaining. If we follow Michael Scriven in defining explanations as descriptions employed to answer "how" questions,[61] then classificatory theories can definitely be seen as explanatory. For instance, the question "how does gametogenesis work?" can be answered by quoting the definition of gametogenesis developed by the Gene Ontology, and one way to answer the question "how does the chicken embryo develop bone structure?" is to list the developmental stages through which bone structure is developed. By contrast, classificatory theories can hardly be seen as explanatory if, by explanation, we require reference to general explanatory principles or to highly general, law-like statements, as required for instance by the deductive-nomological model of explanation. Classificatory theories do not involve law-like, axiomatic statements such as the mathematical equations used in population genetics (e.g., the Hardy-Weinberg law of population equilibrium); they also differ from theories such as evolutionary theory, where a few basic principles provide the tenets for explaining the complex mechanisms of heredity. Strikingly, classificatory theories may display explanatory power but do not set out to work as explanations in the first place. Explanatory power is thus a secondary epistemic virtue in this kind of theorizing. Other epistemic virtues take precedence in shaping this type of formalization: for instance, empirical accuracy, wide intelligibility, and heuristic value for future research.

5.4.4 Providing a Grand Vision Guiding Research. A final characteristic often attributed to scientific theories is their role in providing general frameworks and synthesizing ideas that can inspire and direct future empirical investigations. Such grand visions are typically understood to concern the *contents* of biological knowledge and, more specifically, to encompass a specific view on what constitutes life—as in the case of the Central Dogma in 1960s genetics, for instance, or of evolutionary theory, which despite its various interpretations is typically portrayed as a set of core ideas that brought about a conceptual revolution in the second half of the nineteenth century.[62] I think that classificatory theories do provide a grand vision on the future of biological research but not necessarily by fostering a specific view on what constitutes life. Classificatory theories might well involve ontological commitments on how life evolves and develops. For example, bio-ontologies adhere to the principle of genetic conservation across species, without which the expectation that genetic data on one species may reveal something about another species could not be justified. However, such commitment is only an assumption made for data integration through bio-ontologies to work and does not get developed or challenged within the bio-ontology framework. Indeed, classificatory theories rarely contribute to advancing broad visions on the contents of biological knowledge. Rather, they embody a perspective on *how biological research should be conducted*—a methodological, rather than a conceptual, vision of the study of life. The heuristic role of classificatory theories is thus grounded primarily in methodological commitments, which may carry great epistemic and possibly ontological import depending on how (uncritically or critically) they are embraced by working scientists. Classificatory theory can express knowledge underlying specific research programs, therefore helping identify their core assumptions and reflect on their implications. It codifies what (some) researchers believe about how we should go about studying the world and intervening in it. It is at once a general and a situated approach, which attempts to balance the requirement of comprehensiveness with a deep awareness of the diversity of both biological processes and of the practices used to investigate them. The case of bio-ontologies shows how classificatory theories recognize, and build on, the tension between the wish to make sense of all existing data, which permeates current debates on "data-driven" methods, and the situated nature of data production and interpretation, which makes it so hard for data to travel across contexts and be integrated and reused on the vast scale required by data-intensive efforts. As I already observed in chapter 2, this recognition displays strong ties with the social and cultural world in which such research is embedded, particularly with the tensions between the global and the local, and top-down and bottom-up forms of governance, that feature so prominently at the start of the new millennium. As suggested by the popularity of evolutionary thinking in the nineteenth century, general relativity in the early twentieth century, and the "coding" metaphor in genetics after the Second World War, this ability to latch on to contemporary societal and cultural imaginaries can itself be viewed a key property of grand scientific visions, which are often taken as inspirations for how to understand and intervene in both the biological and the social worlds.

5.5 Theory in Data-Centric Science

I have defended the view that bio-ontologies such as GO play the role of classificatory theories in data-intensive research modes. They consist of an explicitly formulated series of claims about biological phenomena, which is understood in relation to the methods, materials, and instruments used to experiment on those phenomena and routinely adapted to scientific developments. On the one hand, they express the propositional knowledge that is relied on when ordering data for the purpose of further dissemination and analysis, knowledge that expresses current understandings of the significance of data toward understanding biological phenomena and which can be challenged and modified depending on shifts in research contexts. On the other hand, their interpretation is grounded in tacit knowledge of the instruments, models, and protocols characterizing experimental research in biology. Bio-ontologies thus constitute an example of how theory can emerge from classificatory practices in conjunction with experimental know-how. They express what is currently known about biological entities or processes, to further the study of those entities and processes through coordination among research projects and the exchange of relevant data. They formalize knowledge that is taken to be widely assumed yet is usually dispersed across publications and research groups. They need not be universal; rather, they capture the assumptions, interpretations, and practices underlying the successful sharing and reuse of data within very specific contexts at a given moment in time and provide a long-term methodological vision for how society should go about investigating the natural world.

My analysis stands in contrast to a simplistic understanding of data-centric science as juxtaposed to "hypothesis-driven" research—that is, as based on the inductive inference of patterns from datasets leading to the formulation of testable claims without recourse to preconceived hypotheses. I am also highly critical of the hotly debated proposition, publicized to great effect by *Wired Magazine*, that the advent of big data and data-intensive methods is heralding the "end of theory" and the start of a "data-driven" phase of research.[63] As recognized by both champions and critics of data-intensive methods, extracting biologically meaningful inferences from data involves a complex interplay of components, such as reliance on background knowledge, model-based reasoning, and iterative tinkering with materials and instruments.[64] Thus a simplistic opposition between inductive and deductive procedures does not help with understanding the epistemic characteristics of this research mode. Furthermore, we have seen in chapter 3 how data production and interpretation are highly theory-laden processes, which makes it hard to conceive of biological research as driven exclusively (or even primarily) by data; and chapter 4 illustrated how database curators recognize this issue through the provision of metadata. Building on the useful distinction put forward by philosopher Kenneth Waters, I defend the idea that data-centric biology is not theory-*driven* but rather theory-*informed*:[65] it draws on theories without letting them predetermine its ultimate outcomes. The question that needs to be asked about data-centric science is not whether it includes some form of conceptual scaffolding, which it certainly does, but rather whether it uses theories in a way that distinguishes it from other forms of inquiry—and what this tells us about the epistemology of current research. In other words, the question is about how data are being systematized and assembled to yield understanding and what are the key conceptual ingredients and assumptions in that process. This is why studying the role of classificatory theories, and bio-ontologies in particular, is crucial to understanding the epistemic features of data-intensive science as a research mode. It concretely illustrates how the procedures used to extract patterns from data are not purely inductive but rather rely on a sophisticated conceptual structure that is developed to collect, integrate, and retrieve data to make them usable for further analysis.

Viewing bio-ontologies as theories highlights the impact that information technology is having not only on experimental practices but also on scientific reasoning and methods of inquiry. Without computing and the Internet, it would be impossible to structure and regulate bio-ontologies in the ways I described—a fact that makes data-centric science highly dependent on the technological facilities and infrastructures available to researchers, as I discuss in the next chapter. At the same time, emphasizing the conceptual scaffolding embedded in data journeys, and the labor involved in developing, updating, and adapting it to each line of inquiry, constitutes a useful counterpoint to the often-exaggerated role attributed to automation in data-centric science. While digital infrastructure plays an increasingly important role in shaping reasoning and "random" searches, the interpretation of the biological significance of data is grounded in embodied and social knowledge, which help researchers assess and use the theoretical framing provided by bio-ontologies. The situation in which research is carried out— the specific setups, materials, methods—therefore defines the domain of the theory and its interpretation. Database curators are the first to acknowledge that the reuse of data retrieved through automated searches facilitated by bio-ontologies stands in a complex relation to experimental and modeling practices. As I also argued in chapter 4, the analysis of data integrated in silico thanks to classificatory theory provides the best opportunities for discovery when it is used in parallel to other forms of inquiry.

It might seem that viewing bio-ontologies as theories is in contrast with the realist commitments that bio-ontologies seem to express and that are often attributed to them by curators (for instance, when insisting that bio-ontologies represent "the reality captured by the underlying biological science"[66]). I do not think that such conflict exists. Of course bio-ontologies do and should refer to real entities in the world. This does not mean that the knowledge we have of these entities is firm and infallible. Indeed, many biologists prefer to think of their knowledge as tentative rather than conclusive. A leading curator of a model organism database told me that she finds biologists to be put off by the assertive, unambiguous language used by bio-ontologies to express knowledge, and indeed many researchers I interviewed stressed their discomfort with what they perceived as "black and white" formulations. Formulations such as "X is Y" or "A regulates B" are often interpreted by biologists as implying the unwarranted transformation of unverified statistical correlations into well-established causal claims. That absence of qualifiers makes biologists uncomfortable precisely because they know that several of these claims have not been conclusively proven and could well turn out to be mistaken in the future. Thinking of bio-ontologies as theories takes nothing away from the curators' realist commitments to providing the best possible representation of what is currently known about the biological world. Rather, this interpretation stresses that bio-ontologies are as fallible, dynamic, and revisable expressions of biological knowledge, and that is indeed what makes it such an efficient tool for discovery in data-intensive science.[67] Viewing bio-ontologies as theories could thus help to reduce the uneasiness felt by many biologists toward data-intensive methods, by uncovering the conceptual work underlying the development of databases and thus alerting biologists to the pitfalls of an uncritical use of these tools and to the need for constructive critique and expert feedback in order to improve their accuracy and reliability.

More generally, viewing bio-ontologies and other classificatory systems used to systematize data as theories constitutes a way to recognize and uncover the conceptual developments involved in setting up data infrastructures such as databases. STS scholars such as Susan Leigh Star have hitherto called attention to the trouble caused by standards and classification systems when they are "silent"—that is, when they are embedded so deep into the infrastructures that we use that we do not realize the extent to which they are shaping our vision of the world, and we remain unable to challenge them.[68] By conceptualizing classifications as forms of theory, I am instead highlighting the need for extensive and regular critical engagement with the conceptual scaffolding developed to analyze and interpret data. This is what iPlant curators are encouraging when asking their users to provide feedback and what GO curators are attempting to elicit when organizing Content Meetings where various experts can debate how a given term should best be defined. Recognizing the role of classificatory theories as conceptual scaffolds for data journeys is thus important epistemologically, as it can enhance biologists' understanding of how data are assembled and what role concepts play in that process, and institutionally, as it illuminates the conceptual contributions made by curators involved in devising such classifications and makes them accountable for such choices. Theory, in all its forms, can always be made to function as a motor or a hindrance to scientific advancements, depending on the degree of critical awareness with which it is employed in directing research. The fruitfulness of the conceptual scaffolding currently developed to make data travel thus rests on the ability of all stakeholders in biological research to exploit data-intensive methods without forgetting which commitments and constraints they impose on scientific reasoning and practice.

Part Three: Implications for Biology and Philosophy

6

Researching Life in the Digital Age

In the previous chapters, I examined data dissemination practices in data-centric biology and brought my analysis in dialogue with existing philosophical accounts of the dynamics of scientific research. Here I reflect on how these developments affect how at least some life scientists—principally ones drawing on molecular and model organism research as key resources for their work—interact with organisms; how this may influence other subfields, such as those studying evolution and behavior; and how this sits in relation to the history and future trajectories of biology as a whole.

I start by highlighting the epistemic innovations that recent data packaging strategies introduce in relation to how biologists conceptualize and intervene on organisms, particularly in cases where the integration of data coming from a variety of sources is used to challenge and extend existing results. To this aim, I review some examples of data-centric biology that was developed with the help of online databases. I focus on what I view as three types of data integration enacted within contemporary plant biology—each of which emphasizes a specific approach to what organisms are and how they should be studied. These cases illustrate the extent to which data integration via databases can not only yield new biological knowledge but also shape what counts as knowledge in the life sciences in the first place and how biologists organize and equip themselves in order to produce it.

This emphasis on the successes of data-centric biology helps to understand the expectations—and related investments—placed on this form of research over the last decade. At the same time, it risks providing an overly optimistic idea of what the implementation of data mining and related infrastructures can achieve. To counterbalance this impression, I look at some of the dangers attached to the widespread implementation of and reliance on data-centric methods. In previous chapters, I have hinted at several ways in which data journeys can go wrong, with potentially severe consequences for the quality of the knowledge being produced. Section 6.2 garners these issues together and reflects on their potential significance, particularly the possibility that the adoption and progressive entrenchment of particular forms of data packaging into research workflows may foster the marginalization of existing traditions and ways of working within biology. Embedding in digital tools can foster the black-boxing and uncritical treatment of large areas of biological expertise. Knowledge and techniques produced in languages other than English, or research traditions developed far from the central nodes of data dissemination, may thus be ignored and silenced. Moreover, heavy reliance on data circulated through online databases may encourage conservative thinking in biology, rather than spurring new ideas—a clear example of how the very existence and availability of certain kinds of datasets may affect the scientific and social environments into which these data are brought. This brings me to question the representativeness of current success stories in data-centric research with respect to biology as a whole, particularly to fields that rely less on "omics" data and more on data that are as yet too difficult to collect, standardize, and disseminate systematically, such as images, behavioral data, morphological observations, or historical data.

Finally, in section 6.3, I reflect on what these developments signify for how biologists investigate living systems. I review the conditions under which data-centric methods may be expected to open new opportunities for biological research, and I propose that the emphasis on integration and reflexivity characterizing the most sophisticated forms of this approach may well facilitate a processual and dynamic account of biological phenomena, such as that fostered in contemporary metagenomics, epigenetics, systems, and evolutionary-developmental biology. This is most likely when biologists cultivate some understanding of data dissemination strategies and a constructively critical attitude toward their capabilities and limitation. While database curators and computer scientists certainly play a key role in shaping data journeys, their ultimate impact on biology is determined by database users, who therefore need to take some responsibility in the development of these crucial tools—and, given the considerable time and resources involved, should receive relevant support by scientific institutions and funding bodies.

6.1 Varieties of Data Integration, Different Ways to Understand Organisms

Maureen O'Malley and Orkun Soyer have pointed to the scientific work involved in data integration as important and distinct from the work required by other forms of integration, such as methodological and explanatory integration, which have been more successful in captivating the attention of philosophers of science.[1] In this section, I analyze three distinct modes of data integration at work in contemporary biology, which often coexist within the same laboratory but whose competing demands and goals cannot usually be accommodated to an equal extent within any one research project: (1) *interlevel integration*, involving the assembling and interrelation of data documenting different levels of organization within the same species, with the primary aim of improving on existing knowledge of its biology; (2) *cross-species integration*, involving the comparison and co-construction of data on different species, again with the primary aim of widening existing biological knowledge; and (3) *translational integration*, involving the use of data from a wide variety of sources to devise new forms of intervention on organisms, ultimately aimed at improving human health. As examples for each of these forms of integration, I use three case studies from contemporary plant science: (1) the research activities centered around data gathered on model organism *Arabidopsis thaliana*; (2) the efforts to integrate *Arabidopsis* data with data gathered on the perennial crop family *Miscanthus*; and (3) current investigations of the biology of *Phytophthora ramorum*, a plant parasite that has been wreaking havoc in the forests of the southwestern United Kingdom, where I live, since the early 2000s.

There are several reasons for my focus on plant biology. First, despite its enormous scientific and social importance, this is a relatively small area of research—especially in comparison to biomedicine—and has received little attention within the philosophy of science.[2] It is also heavily funded by governmental agencies, particularly when it comes to research relating to molecular and genomic aspects of plants—a fact that enhanced opportunities for plant scientists to work on foundational questions, as well as pushing them to join forces and collaborate in order to attract the attention of sponsors and make the best of limited resources.[3] Plant science has indeed been open to generalist and interdisciplinary thinking throughout its history, as even the most reductionist plant scientists tend to be greatly interested in the relations between molecular biology, cellular mechanisms, developmental biology, ecology, and evolution.[4] Further, and similarly to the case of model organism research, prepublication exchanges have been strongly encouraged within this community, particularly at the time of the introduction of molecular analysis on plants. The latter was spearheaded by a group of charismatic individuals who explicitly aimed to advance knowledge through open collaboration, thus effectively predating today's Open Science movement[5]. These factors made the community of plant scientists into a relatively more cohesive and collaborative one than more powerful, socially visible, well-funded, and tightly structured fields focusing on animal models, such as cancer research or immunology.[6] Given this background, it is not surprising that plant science has produced some of the best available resources for scientific data management and integration to date. Plant scientists' interest in working together, and thus in finding efficient ways to assemble and disseminate their resources and results, long precedes the advent of digital technologies for data sharing, and many of these scientists were quick to seize the potential provided by those technologies to help them in their integrative efforts. As a result, plant science possesses some of the most sophisticated databases and modeling tools in biology.[7] Its contributions to systems biology are substantial, particularly in fostering the development of digital organisms (i.e., the use of models and simulations to integrate qualitative data, in order to predict organismal behavior and traits in relation to the environment).

Yet another characteristic of plant science makes it a fruitful terrain on which to study data integration as a key process in producing new knowledge. Plant science produces results of direct interest to several sections of society, including farmers, forestry management, landowners, florists and gardeners, the food industry, and the energy industry (through the production of first- and second-generation biofuels), social movements concerned about genetically modified foods, sustainable farming and population growth, breeders of new plant varieties for agricultural or decorative use, and of course national and international government agencies. Representatives of these groups can be and sometimes are called to participate in the development and planning of scientific projects, thus contributing to choice of goals, opportunities, and constraints associated to ongoing research. In significant ways, direct contributions to scientific research by nonscientists make a difference not only to the goals ultimately served by science but also to its practice, methods, and results, including what strategies are used to share and integrate data and what comes to count as new scientific knowledge arising from such integration. As I will illustrate, recognizing the differences in the degrees to which scientific inquiry is brought in contact with other sections of society involves challenging the internalistic view of scientific knowledge that is still favored by many philosophers of science, thus bringing my arguments to bear on recent debates around the social relevance of philosophy of science.[8] Going beyond the view of science as aiming solely to acquire true knowledge of the world may seem a long shot when starting from an analysis of different forms of data integration in contemporary plant science; and yet, as I argue in what follows, looking at processes of integration "in action" immediately points to important differences in the types and sources of data being integrated, the integrative processes themselves, and the forms of knowledge obtained as a result of integration.[9]

6.1.1 Model Organisms as Reference Points: Interlevel Integration. In chapter 1, I illustrated how databases have played pivotal roles in defining the epistemic roles of model organisms in contemporary biology. They greatly facilitated the fulfilment of the premises on which model organism communities were founded: a collaborative ethos and willingness to share results so as to understand organisms as "wholes," a related need to exchange materials and specimens, and the resulting efforts to standardize nomenclature and experimental protocols in order to make exchanges as seamless and global as possible. Occupying this niche allowed community databases to demonstrate how fruitful collaboration around and across model organisms could be. This process, which was a continuation of the history of model organism research since the turn of the twentieth century, resulted in an acceleration of the building of these communities and their expansion into a global research force (and a considerable share of the funding available for biological and even biomedical research).

On the one hand, model organism databases thus reinforced the power of popular model organisms over other organisms with less well-organized communities. Indeed, the popularity of particular model organisms has tended to grow incrementally with the scale and organization of their community databases—a principle readily recognized by biologists wishing to promote new organisms as "biology's next top models," according to whom obtaining funding to build a community database is a crucial step in the process,[10] and by researchers wishing to highlight the usefulness of model organism research for the future study of human disease[11] and evolutionary developmental biology.[12] On the other hand, however, it is important to note how the expansion of community databases has shifted attention away from research on single species and toward comparative, cross-species research that takes account of environmental variability. Model organism databases are taken as reference points for the investigation of other species in the same family or kingdom about which less is known and for the study of how environmental conditions affect the expression of specific traits. For instance, The Arabidopsis Information Resource (TAIR) continues to be a popular tool among researchers focused on crops or trees, because it provides a reference point for how specific processes (such as vernalization, cell metabolism, or root development) might work and which genes and biochemical pathways might be involved. WormBase is also increasingly used by researchers working on nematodes other than C. *elegans*, such as those that are significant agricultural or human parasites.

Hence databases have made essential contributions to the development of an understanding of model organisms as tools that permit *comparison* across species. Research on model organisms began in part to provide reference points for such comparison, and the use of databases has enhanced their capabilities to act as such reference points. The strength of model organisms as comparative tools lies in their capacities to represent specific groups of organisms, as well as to enable cross-disciplinary research programs exploring several different aspects of their biology, with the ultimate goal of reaching an integrative understanding of the organisms as intact wholes. Data infrastructure, in the form of online databases and thus of the communities, specimen collections, and information to which they provide access, is the platform over which model organisms can now define themselves as comparative, representative, and integrative tools. Without data infrastructure, the exchange of information about model organisms and their use for comparative purposes would be impossible to realize on the appropriate scale. By bringing results, people, and specimens together using infrastructure, community databases have thus come to play a crucial role in defining what counts as knowledge of organisms in the postgenomic era.

Model organism research has been immensely successful within plant science, where it considerably enhanced the understanding of key processes such as photosynthesis, flowering, and root development.[13] Understanding these processes requires an interdisciplinary approach comprising several levels of organization, from the molecular to the developmental and morphological.[14] *Arabidopsis* provided a relatively simple organism on which integration across these levels could be tried out under the controlled conditions of a laboratory setting. The use of *Arabidopsis* has been highly controversial within plant science at large, with scientists specializing on the study of other plants and/or plant ecology complaining that focusing on *Arabidopsis* took resources away from the study of plant biodiversity, evolution, and the relation between plants and their environment.[15] At the same time, considering *Arabidopsis* in relative isolation from its natural environments and other plants has been successful in generating important insights about its inner mechanisms, particularly at the molecular and cellular levels. For instance, the detailed mechanistic explanations of photosynthesis achieved to date, and the resulting ability of scientists to manipulate starches and light conditions to favor plant growth, are largely due to successful attempts to bring the study of enzymes and other proteins involved in photosynthesis (which involves the analysis of molecular interactions within the cell nucleus) in relation with the study of metabolism (which involves the cellular level of analysis, since it focuses on posttranslational processes outside the nucleus). The integration of these two levels of analysis is fraught with difficulties, since the evaluation of data about DNA molecules (as provided by genome analysis) needs to take account of their actual behavior within and interactions with the complex and dynamic environment of the cell, which makes it extremely difficult to model metabolic pathways.[16]

This is an excellent illustration of interlevel integration as aiming to understand organisms as complex entities, by combining data coming from different branches of biology in order to obtain holistic, interdisciplinary knowledge that cuts across levels of organization of the same organism.[17] This case also instantiates the key role played by databases and curatorial activities in achieving interlevel integration, as the development of centralized depositories for data has been central to the success of model organism research.[18] Since its inception, TAIR has been heavily engaged in facilitating interlevel data integration, particularly through the development of software and modeling tools that enable users to combine and visualize datasets acquired on two or more levels or organization. Tools such as AraCyc and MetaCyc, for instance, have enabled researchers to combine and visualize genomic, transcriptomic, and metabolic data as a single body of information. This has made it possible to integrate data generated from the molecular and the cellular levels of organization, thus enabling researchers to visualize and study specific metabolic pathways. TAIR curators have also devoted much attention to developing metadata that help researchers working at a specific level (e.g., cellular) to assess and interpret data gathered at another level (e.g., molecular). Last but not least, TAIR curators have endeavored to collaborate with researchers from all corners of plant science to generate bio-ontologies that could adequately describe the biological objects and processes currently under investigation, such as Gene Ontology (GO) and Plant Ontology (PO) (which I discussed in chapters 1, 4, and 5).

It is important to stress that interlevel research on *Arabidopsis*, as on many other popular model organisms, was driven strongly by the scientific community, with the support of funding bodies such as the National Science Foundation, but with little influence from other parts of society that have stakes in plant science—such as, for instance, agricultural research, farmers, and industrial breeders.[19] Attempting to integrate data resulting from different strands of plant research was seen as requiring expert consultations within the plant science community, aimed at resolving technical and conceptual problems in an effort to acquire an improved understanding of *Arabidopsis* biology. The very idea of using data from the same model organism to achieve interlevel integration exemplifies the image of science often celebrated within philosophy of science circles, as well as many popular accounts of discovery: these are scientists who wish to bring together their results within expert circles that are largely separate from other sections of society; and that this is done to acquire a more accurate and truthful understanding of biological processes, resulting in the articulation of reliable explanations of those processes (and related forms of intervention).[20] Indeed, interlevel integration is heavily concerned with the methodological and conceptual challenges deriving from the attempt to collaborate across disciplines, such as the effort to standardize the propositional, embodied, and social knowledge produced by each community. In chapter 1, we have seen how iPlant curators continue to engage in extensive consultations with plant scientists working on several levels of biological organization, so as to make sure that ensure that data integration tools are set up in ways agreeable to and compatible with research at many levels of analysis.

These efforts are reminiscent of the challenge of communicating propositional knowledge across different scientific groups, which many philosophers of science have focused on when reflecting on scientific integration.[21] As we saw in chapter 4, focusing on data integration, rather than on the integration of explanations, models, and theories addressed by these authors, helps highlight the importance of communicating embodied knowledge in order to achieve new insights (a factor that tends to be overlooked in literature focused on explanations and conceptual structures). The focus on data also helps stress the diversity of the epistemic goals and priorities driving integration, as well as the distinct forms of knowledge that may be achieved through pursuing these goals. To this end, I now discuss two forms of integration that operate differently from interlevel integration and whose results are distinct from the knowledge of plant biology acquired in this case.

6.1.2 From Model Organisms to Situated Biodiversity: Cross-Species Integration.

In cross-species integration, scientists place more emphasis on comparing data available on different species, and using such comparisons as a springboard for new discoveries, rather than on integrating data across levels of organization of the same species.[22] Consider current research on grass species *Miscanthus giganteus* (figure 7). *Miscanthus* is a perennial

FIGURE 7 A stand of *Miscanthus x giganteus* from the German company Sieverdingbeck, October 2011. (Hamsterdancer, Wikimedia Commons, October 2015, https://commons.wikimedia.org/wiki/File: Miscanthus_Bestand.JPG.)

crop, which means that it can be cultivated in all seasons without interruptions to the production chain. It grows fast and tall, thus guaranteeing a high yield, and it grows easily on marginal land. These characteristics have made it a good candidate as a source of bioethanol, particularly because it poses less of a threat to food production than other popular sources of biofuels such as corn (whose cultivation for the purposes of biofuel production has taken big chunks of land in the United States away from agriculture, which is deemed to have affected the availability and price of agricultural produce worldwide).[23]

The potential of *Miscanthus* as a source for biofuels is one of the factors that first spurred scientific research on this plant. And indeed, such research is ultimately aimed to engineer *Miscanthus* so that its growing season is extended (by manipulating early season vigor and senescence) and its light intake is optimized (modify architecture via several sprawling stems or increase the stem height and number). In this broader sense, research on *Miscanthus* is a good example of research aimed at developing techniques for intervening in the world and ultimately for improving human life. However, there is at least another reason *Miscanthus* has become an important organism in contemporary plant science, which has little to do with its energy output. This is the opportunity to efficiently cross-reference the study of *Miscanthus* with research on *Arabidopsis*.

On the one hand, *Arabidopsis* provides the perfect platform on which to conduct exploratory experiments, given how much scientists already know about that system (thanks to interlevel integrative efforts) and the extensive infrastructure, standards for collecting data and metadata, and modeling tools already available on it (e.g., as incorporated in TAIR and iPlant). On the other hand, *Miscanthus* provides a good test case for ideas first developed with reference to *Arabidopsis*, whose value for other species researchers have yet to explore. Many researchers trained on *Arabidopsis* biology have thus switched to comparative research on these two systems, which has hitherto proved very productive: many experiments needed to acquire knowledge about molecular pathways relating to abiotic stress can be more easily carried out on *Arabidopsis* than on *Miscanthus*; data collection and integration on *Miscanthus* itself is facilitated by the standards, repositories, and curatorial techniques already developed for *Arabidopsis*; and new data types, such as data about how *Miscanthus* behaves in the field (e.g., its water intake), can be usefully integrated with data about *Arabidopsis* metabolism, resulting in new knowledge about how plants produce energy in both species. This research requires more than simply the transfer of knowledge from one plant to the other. To obtain new knowledge, plant scientists need to iteratively move between the two species, compare results, and integrate data at every step of the way. In other words, this research requires genuine integration between results obtained on *Miscanthus* and *Arabidopsis*. For instance, the consultation of TAIR data on *Arabidopsis* genes that regulate floral transition has been a crucial impetus for research on flowering time in *Miscanthus*, since those data provided *Miscanthus* researchers with a starting point for investigating the regulatory mechanisms for this process;[24] and the subsequent findings on the susceptibility of *Miscanthus* flowering to temperature and geographical location are feeding back into the study of flowering time in *Arabidopsis*.[25]

Perhaps the most important distinctive feature of cross-species integration is that it fosters studies of organismal variation and biodiversity in relation to the environment, with the aim of understanding organisms as relational entities, rather than as complex—yet self-contained—wholes (as in the case of interlevel integration). This is because as soon as similarities and differences between species become the focus of research, plant researchers need to identify at least some of the reasons for those similarities and differences, which unavoidably involves considering their evolutionary origins and/or the environmental conditions in which they develop. Hence, like interlevel integration, cross-species integration may be construed as aiming to develop new scientific knowledge of biological entities. However, the way in which it proposes to expand the realm of existing knowledge is not necessarily by extending the range of interlevel explanations available, but rather by extending the range of organisms to which these explanations may apply. Indeed, while researchers can and often do pursue both interlevel and cross-species integration at the same time, it is also possible to achieve cross-species integration without necessarily fostering interlevel integration. This is the case, for instance, when using comparisons of data about flowering time between *Arabidopsis* and *Miscanthus* to explore the respective responses of the two plants to temperature; such cross-species comparison can eventually be used to foster interlevel understanding of flowering that integrates molecular, cellular, and physiological insights, but this is not necessary in order for cross-species integration to be regarded as an important achievement in its own right.

Furthermore, cross-species integration poses a different set of challenges from interlevel integration, whose resolution can easily constitute the sole research focus of a research project. It requires accumulating data that are specifically relevant for the purposes of comparison (for instance, by making sure that data obtained on *Miscanthus* are generated with tools and on materials similar to the ones available on *Arabidopsis*, so as to make comparison tenable), as well as developing infrastructure, algorithms, and models that enable researchers to usefully visualize and compare such data. Thus in our example, TAIR provides a key reference point, but it is not sufficient as a data infrastructure for such a project, for the simple reason that it focuses on *Arabidopsis* data alone. Indeed, the difficulties of using TAIR for cross-species integration have become so pronounced and visible within the plant science community that TAIR itself is now complemented by another *Arabidopsis*-centered database, Araport, in order to secure its future compatibility with *both* interlevel and cross-species analysis.[26] This task is made even more difficult by the terminological, conceptual, and methodological differences between communities working on different organisms, as well as differences in perceptions of what counts as good evidence and the degree to which specific traits are conserved across species through their evolutionary history. These differences need to be clearly signaled when constructing databases that include and integrate data acquired on different organisms. GO, for instance, is now used extensively as a platform for the integration of gene products data across species[27] and exemplifies the difficulties and controversies involved in rising to this challenge.[28] Even the comparison of different genome sequences, which should be among the easiest to accomplish given the highly automated and standardized production of this type of data, is fraught with problems.[29] In this context, norms such as the principle of genetic conservation, by which scientists see regions of the genome that are highly conserved across species as potentially linked to important functions (since less relevant regions are assumed to have been selected away through the evolutionary process), matter over and above the norms of validity and accuracy used to achieve interlevel integration.

In conclusion, the increasing emphasis on cross-species integration can be seen as complementary to, and yet separate from, interlevel integration. The two forms of data integration are clearly interconnected. I have illustrated how the interlevel integration achieved for *Arabidopsis* through model organism biology provides an important reference for cross-species integration in several areas of plant science. Interlevel integration of data within one species are often the starting point for cross-species investigation and for the integration of data about the same process as it manifests itself in different species. However, this does not necessarily mean that cross-species integration presupposes interlevel integration as a matter of principle, or in all cases. Further, these two forms of integration raise different epistemic challenges, which do not have to be addressed within the same research project; and, most importantly for my present purposes, they require different sets of data and infrastructures—as illustrated by the practical difficulties in using TAIR, whose primary focus is interlevel integration in *Arabidopsis*, to study other plant species such as *Miscanthus*.

6.1.3 When Social Agendas Guide Research: Translational Integration. Research on *Miscanthus* could be seen as exemplifying research that has been targeted and structured to serve societal goals—in this case, the sustainable production of biofuels. However, while the goals set by funding agencies and industry have been crucial to the choice and funding of *Miscanthus* as an experimental organism, plant scientists engaged in *Miscanthus* research have not, at least until recently, worried much about how the plants that they are engineering could actually be transformed into biofuel, whether that process would be particularly sustainable and economical, and how those "downstream" considerations might affect "upstream" research. This set of consideration has largely been left to politicians and industry analysis, while plant scientists focus on the task of achieving new knowledge of *Miscanthus* biology. In other words, scientists and curators focusing on cross-species integration are primarily focused on producing knowledge about plant biology that is more accurate and all-encompassing than that already available to them. Their expectation is that knowledge produced in this way will eventually inform the mass engineering of *Miscanthus* plants, thus creating biomass from which bioethanol could be efficiently extracted. This is a reasonable expectation, and the knowledge obtained from *Miscanthus* research will undoubtedly inform biofuel production in the future. However, other research programs in plant science are explicitly planned and shaped to serve societal needs *even before* they improve on existing scientific knowledge of the organisms involved, thus de-emphasizing the production of new biological knowledge in favor of producing strategies for managing and manipulating organisms and environments so that they support human survival and well-being in the long term.

Consider for instance research on plant pathogens, which are becoming a serious threat to ecosystems and agriculture worldwide because of global trade and travel that facilitate the dispersion of parasites well beyond their natural reach.[30] Dealing with plant pathogens that are new to a given territory is a matter of urgency, since targeted interventions need to be devised before the pathogen creates much damage. Scientific research is a key contributor, as these pathogens are often relatively unknown within the scientific literature and are anyhow interacting with a whole new ecosystem, often with unprecedented results. In February 2012, I participated in a workshop organized at the University of Exeter to discuss how plant science can help to suppress an infestation of *Phytophthora ramorum*, a plant parasite that landed in the southwestern United Kingdom in the early 2000s and has been ravaging the forests of Devon ever since. The infestation had become particularly worrisome in 2009, when it started to affect large chunks of the local population of larches (figure 8). The workshop brought together representatives of relevant plant biology and data curation conducted at several research institutes in the United Kingdom; the UK Forestry Commission; the Food and Environment Research Agency; private landowners; social scientists; and representatives of other governmental agencies, such as the Biotechnology and Biological Sciences Research Council.

At the start of the workshop, it was made clear that there are several alternative ways to tackle the *Phytophthora* infestation, including burning the affected areas, using pesticides, cutting down the trees, letting trees live and introducing predators, making affected areas inaccessible to humans, or simply letting the infection run its course. A focus of debate was then to determine which scientific approach would provide empirical grounds to choose an effective course of action among all the possible interventions. Acquiring novel understandings of the biology of *Phytophthora* was obviously important in this respect; but it was not the primary goal of the meeting, and it was made clear that choices concerning which research approach would be privileged in the short term should not be based on the long-term usefulness of that approach in providing new biological insights. This was particularly relevant in selecting strategies for data collection and

FIGURE 8 Hillside in Big Sur, California, devastated by *Phytophthora ramorum* infestation. While practically all woody plants in such forests are susceptible (Rizzo et al., *"Phytophthora ramorum* and Sudden Oak Death in California 1"*)*, American oaks like tanoaks are particularly vulnerable. By contrast, no significant damage has been observed in European oaks, while in Britain the disease has decimated larch species. (Rocio Nadat, Wikimedia Commons, October 2015, https://en.wikipedia.org/wiki/File:Sudden _oak_death_IMG_0223.JPG.)

types of data to be privileged in further analysis. For instance, whole genome sequencing was agreed to be an excellent starting point for a traditional research program seeking to understand the biology of *Phytophthora* through interlevel and cross-species integration, especially since data could be compared (through online databases) to data generated on other strains of *Phytophthora* by European and North American labs. However, many participants questioned the efficiency of this strategy in providing quickly genetic markers for *Phytophthora ramorum*, which could be of immediate use to combat the infestation. It was argued that focusing genomic research on more specific parts of the genome, such as loci already known to be linked to pathogenic traits, would provide a way for the Forestry Commission to test trees in areas not yet affected and determine immediately whether the infestation was spreading (the merits and drawbacks of using diagnostic based on polymerase chain reaction [PCR] technology were debated at length). Further, much debate surrounded the possible ecological, economic, and societal implications of each mode of intervention under consideration and the science related to it. Biological research was thus not the sole empirical ground to assess the quality and effectiveness of an intervention; other factors included local ecology, touristic value of the areas, and the economic value of the wood being cut down (i.e., factors that include the environmental considerations that I signaled as central to the cross-species approach, as well as economic and social elements that cross-species integration would not regard as relevant). Only through such an overall assessment could participants and scientists determine the overall sustainability of the research program that was being planned.

Notably, each participant contributed not only a specific perspective on what the priorities are in dealing with *Phytophthora* but also their own datasets for integration with the molecular and phenotypic data to be gathered by plant scientists. These included data of great relevance to scientific research, although they were collected for purposes other than the study of *Phytophthora* biology: for instance, geographical data about the spread of the infestation, which were gathered by Forest Research (the research arm of the Forestry Commission) in the course of aerial surveillance and picked up by mathematical modelers to help predict future spread patterns, and photographs of affected trees in several areas, collected by the Forestry Commission and local landowners and seized upon by plant pathologists at the James Hutton Institute as evidence for plant responses to biotic stress. Acquiring access to those data constitutes an achievement in itself for plant scientists, since some of the stakeholders involved are more willing to disseminate their data than others. For instance, the Forestry Commission is more reluctant to share data than plant scientists working at the University of Exeter, for whom contributing to online sequence repositories such as the Sequence Read Archive or GenBank is a routine part of research. Further, scientists at the meeting were not sure about which existing online database, or combinations of databases, would best serve the desired integrative efforts. One obvious candidate would be PathoPlant, a database explicitly devoted to data on plant-pathogen interaction, but its use was not explicitly discussed at the meeting I attended, perhaps because it was not clear to participants that such a database would serve their immediate research goals.[31]

Indeed, plant scientists involved in the meeting at the University of Exeter found themselves negotiating with stakeholders, some of whom are also arguably involved in scientific research (such as biologists working for the Forestry Commission), whose main aim was not the production of new insights on *Phytophthora* biology but rather the achievement of a reliable body of evidence that would help with deciding how to tackle *Phytophthora* infestations. This negotiation, which is the key feature of this type of integrative effort, is not easy, especially given the tendency of plant scientists to reach for interlevel and/or cross-species integration too. Thus plant scientists at the meeting strongly advocated the expansion of research on *Phytophthora ramorum* into a long-term program that would investigate the relative virulence of the pathogen on different hosts (which would involve detailed studies of the hosts—tree species—as well), assemble whole genome data on all available and emerging strains, and investigate the mechanisms that trigger virulence. All these research programs, which clearly involve interlevel and cross-species data integration, would provide knowledge about the biology of *Phytophthora* that scientists view as crucial to developing better interventions. However, these programs would require substantial funding and considerable time in order to yield results, and scientists were pushed by other parties to articulate more fully how the systematic whole genome sequencing of *Phytophthora* strains would eventually lead to effective interventions on the infection. In particular, it was argued that although the cross-species and interlevel integration acquired through these approaches was desirable, it was not necessary in order to facilitate decisions on how to eradicate *Phytophthora*. For example, PCR-based diagnostics, though arguably useless to the pursuit of a better understanding of the biology of *Phytophthora* and its hosts, might work perfectly well for the purposes of diagnosing infection.

Negotiations among molecular biologists, scientists working at Forest Research, and other stakeholders are still ongoing at the time of writing, and engagement in these discussions is generating a shared research program, part of which will involve the development of a database that fosters the integration of data relevant to the study the virulence and potential environmental impact of *Phytophthora*. This case nicely exemplifies the characteristics of translational integration, which privileges the achievement of improvements to human health, for instance through targeted interventions on the environment and the use of existing resources, over the production of new scientific knowledge for its own sake.[32] I take the term "translational" from current policy discussions of the importance of making scientific research useful to wider society, as instigated for instance by the National Institutes of Health in the early 2000s. However, I do not subscribe to the linear trajectory of research from "basic" to "applied" that is often used within such policy discussions. Rather, I use the category of "translation" to focus on specific ways in which scientists frame their research so as to respond to a social challenge. In the case I considered, scientists aim to produce new forms of intervention that are targeted to the situation at hand. This is not in itself sufficient to differentiate translational integration from interlevel and cross-species integration. As I argued in chapter 4, learning to intervene in the world, particularly to manipulate organisms in the case of biology, is part and parcel of scientific research and is inseparable from the process of acquiring new knowledge about the world. There is thus no clear epistemic distinction between "making" and "understanding," and many scientists develop new types of experimental interventions as a way to acquire new knowledge, and vice versa.

What I think makes translational data integration distinct from the other two modes is the strong commitment to producing results that affect (and hopefully improve) human health, which involves the development of research strategies and methods that are distinct from the ones employed to achieve interlevel and cross-species integration. My definition of translation is therefore narrower than the definition provided by O'Malley and Karola Stotz, according to which translational research consists of "the capacity to transfer interventions from context to context during the pluralistic investigation of a system."[33] I agree with them that translational research involves such movement of knowledge, but I also think that these transfers can be geared toward satisfying a variety of agendas, several of which are not primarily concerned with how scientific knowledge affects society. Many parts of biological research inherit and refashion techniques for intervening on organisms, without necessarily aiming to produce socially valuable results in the short term. All research has the potential to ultimately improve human health, and yet some parts of science are not explicitly conducted to foster this goal in the short term (which is, incidentally, a very good thing, both because the potential social benefits of science are unpredictable, and because the social agenda for what counts as beneficial to humanity changes with time and across domains). I see the extent to which a group of scientists explicitly subscribes to the agenda of social change—and shapes its research accordingly—as marking the difference between more "foundational" scientific research and translational endeavors. I therefore would not agree with O'Malley and Stotz when they conclude that translation is involved whenever techniques for intervention are transferred from one scientific context to another. In their definition, translation is involved, potentially to the same degree, in all three modes of integration that I consider here; while in my analysis, translation becomes a primary concern, with important consequences for how research is conducted and with which outcomes, when scientists commit to fulfilling specific social roles in the short term.

A key implication of the commitment to improving human health is that scientists engaged in translational data integration need to pay attention to the *sustainability* of their research program—not only in the narrow sense of worrying about its financial viability but also in the broader sense of considering the potential environmental and social impact of the understanding of organisms that they will generate. In practice, this typically involves engaging directly with contexts of production/use, so as to be able to assess the "downstream" applicability of specific research strategies and prospective results. Crucially, biologists do not possess the right expertise to determine, by themselves, what counts as "sustainable" research outcomes. This is why they need to collaborate with scientists in industry, state agencies, and social scientists, among others, as it is through such engagements that scientists determine what constitutes "human health" and how to improve it in the case at hand. Indeed, the "social agenda" for translational research cannot be fixed, for the simple reason that it depends heavily on the ever-changing viewpoints and needs of the many stakeholders involved. Scientists who choose to take time to discuss the goal and outcomes of their research with relevant parties outside the scientific world, and tailor their own research, tools, and methods to fit those discussions, are investing a significant amount of their resources on producing results that might not be revolutionary in terms of their conceptual contribution to existing biological knowledge (though they may prove to be such!), but rather are primarily meant to serve a wider social agenda.

Researchers involved in these exchanges are often also forced to compromise on their own views of what would constitute a productive research strategy and attractive research findings, in order to accommodate requirements and suggestions by other parties interested in achieving social, rather than scientific, goals. In particular, prioritizing the achievement of sustainable and efficient intervention (where what counts as sustainable and efficient is agreed on among several different parties) over the acquisition of biological knowledge has important consequences for processes of data integration, such as the choice of relevant data and the speed with which data need to be collected and interpreted.[34]

6.1.4 The Plurality of Data-Centric Knowledge Production. Paying attention to the differences and interplay between the modes of data integration discussed above exemplifies the challenges of making data usable to the broader scientific community; the large amount of conceptual and material scaffolding needed to transform data available online into new scientific knowledge; and the different forms of knowledge that may result from processes of data integration, depending on which communities, infrastructures, and institutions are involved in scientific research. My analysis also underscores the importance of considering the whole spectrum of research activities, including so-called applied research carried out by industry or governmental agencies, to develop and improve current philosophical understandings of scientific epistemology. Any one component of scientific research (whether data, models, or explanations) can potentially contribute to enhancing opportunities for biomedical as well as environmental and agricultural interventions, all of which are of potential value to the preservation and improvement of human health. Taking the potential social impact of scientific research into account has implications for how the philosophy of science makes sense of the different strategies that scientists may develop when pursuing a research program.

The typology proposed here is not meant to be exhaustive of all the ways in which data integration—and the related data journeys—can stimulate knowledge production in plant research. These forms of integration are also not meant to be mutually exclusive, and very often they happen alongside each other, and in dialog with each other, within the same scientific laboratory (as my examples have shown). Most research groups that I have come across are interested and involved in all three types of data integration and thus contribute to understanding organisms in a variety of different ways. There are good reasons for this state of affairs. The interplay among those three modes is often crucial to the scientific success of a lab, especially at a time when both scientific excellence and the social impact of research are highly valued by funding bodies across the globe and heated discussions surround the choice of metrics to assess how scientists fare on these two counts. The development of standards enabling the integration of data across species has been a key step in the development of model organism databases, thus signaling the willingness of researchers to move beyond interlevel integration and toward cross-species comparisons. Similarly, both interlevel and cross-species integration routinely generate results on which new forms of translational integration can be built—a situation that motivates much of the current investment in big data by corporations and governments. The very case of *Miscanthus* might work in this way if plant scientists actively engage in the search for efficient and sustainable ways to downstream the production of bioethanol and use these interactions to inform their own bioengineering practices (which seems to be exactly what plant scientists are starting to do, for instance through initiatives such as the UK Plant Science Federation). There are even cases where all three types of integration are attempted simultaneously, such as the effort to combine transcriptomic data obtained from large groups of plants (cross-species integration) with metabolic profiling, functional genomics, and systems biology approaches (interlevel integration) so as to reveal "entire pathways for medicinal products," in ways that promise to revolutionize drug discovery and thus provide a good instance of translational integration.[35]

Given the multiple and complex interrelation between the three forms of integration that I have identified, one might wonder why it matters to distinguish them at all. The reason has to do with improving existing philosophical understandings of scientific practices and of the temporality and constraints within which research is carried out. Even if these three forms of integration are intertwined in the overall vision of what science is supposed to achieve for humanity, and in the overall trajectory of any one specific research group, their distinctive aims, methods, strategies, and norms require that they are taken up to different degrees at any one point in time. All the plant scientists whose work I have discussed here are interested in using data to understand the biology of specific plants as an integrated whole; compare different species so as to reach as encompassing an understanding of the plant kingdom as possible (including its evolution, inner diversity and environmental role); *and* address key challenges to human life in the twenty-first century, such as climate change, urbanization, and population increase.[36] Yet *they are unable to pursue all these goals in equal measure at the same time*;[37] and the choices that they make when considering which data to view as most relevant to their research, how to integrate those data, and which expertise to involve in that process will be crucial factors in determining which form of knowledge they prioritize as the primary outcome of their efforts. In other words: the pursuit of different forms of data integration, and the trajectories thus taken by data journeys, give rise to different forms of scientific knowledge, whose value and content shifts in relation to the goals, expertise, and methods involved in each research project.

6.2 The Impact of Data Centrism: Dangers and Exclusions

Through the analysis of research practices given in the previous section, I have shown how the choice of the infrastructure and standards used to disseminate and integrate data affect decisions about which research goals to pursue and thus which forms of knowledge to achieve; and, at the same time, how prioritizing specific goals over others might lead to structuring data journeys and integration, and the infrastructures and standards used to that effect, in different ways. Considering the procedures and standards developed to facilitate data integration also provides important clues about the norms, practices, and implications of integrative processes and the epistemic significance of the social and institutional contexts in which such efforts take place—particularly when it comes to determining who gets to contribute to data journeys and in which capacity. In this section, I discuss the future prospects of data-centric biology, particularly of the expansion of data journeys through the packaging strategies, conceptual scaffolds, and infrastructures documented in previous chapters, with reference to the embedding of those practices into specific institutional, material, and social realities. This brings me to highlight some of the concerns and dangers associated with the forms of data-centric research that I discussed in this book.

I start by identifying four major obstacles hampering the material realization of the extensive data travel required by data-centric research.

The first is the availability of adequate *funding*. We have seen how data packaging activities require large investments in manual curation in order to adequately facilitate data journeys and data reuse within new contexts. However, funding for database curation remains relatively scarce, especially when compared to the investments made by public and private institutions in other research activities. This is true even in areas as successful as model organism biology, where the best stocked and curated databases are those concerned with few, highly standardized data types (such as sequencing, transcriptomics, and, to some extent, proteomics), while curators are still struggling to incorporate more labor-intensive data such as those used in metabolomics, cell biology, physiology, morphology, pathology, and environmental science.[38] The problem becomes more severe for research areas that are less valued by funding bodies, and thus receive less support, or fields in which data production remains extremely expensive. In both cases, most biologists prefer to devote their resources to data generation and analysis rather than to activities such as the public dissemination of their hard-won results or the provision of feedback to existing databases, which are less likely to bring immediate rewards.

A second obstacle is the lack of *engagement* of biological researchers in data curation activities. I have stressed how the development of adequate data packaging requires the support and cooperation of the broader biological community. However, such support is strongly limited by the credit regimes to which biologists are subject, which typically do not valorize donation to databases and participation in their curation. Hence many scientists, particularly those subject to severe financial and social pressures, perceive data handling activities as an inexcusable waste of time, despite being aware of their scientific importance. As I discussed in chapter 2, the most substantial engagement with data travel tends to come from groups working in powerful and well-endowed institutions and on research areas widely recognized as "cutting edge" and incentivized by funding bodies. This kind of setting makes it more likely for biologists to be able to devote time and resources to data donation and curation. Researchers working in prestigious institutions are also more likely to capitalize on their visibility and international links, thus valorizing data curation activities as providing new platforms for research collaboration and exchanges. Thus, and contrary to claims made about the potentially democratizing power of widespread data dissemination, the circumstances required for researchers to engage in data journeys show that the digital divide is alive and well in data-centric biology.

This becomes even clearer when considering the remaining two obstacles to data travel on my list. The third is the limited availability of necessary *infrastructure*, such as a functioning broadband network, adequate computing facilities, and a steady supply of electricity. Researchers' ability to access online databases, and thus to retrieve and evaluate data therein, depends on such seemingly basic infrastructure being reliably in place; and yet these facilities are not easily available to researchers working outside of large cities or highly urbanized areas, such as people based in rural parts of the world (e.g., Wales or the Scottish highlands in the United Kingdom) and/or in so-called developing regions (e.g., sub-Saharan Africa or Bangladesh). Despite dramatic improvements in broadband coverage and the availability of computing tools, recent research has found the digital divide to be over-whelmingly increasing.[39] Furthermore, other infrastructural problems—such as the lack of reliable public transport, courier systems, and refriger-ating facilities—can make the retrieval and maintenance of experimental materials and equipment a real challenge, resulting in much less time and willingness to engage with researchers outside one's immediate community and thus to donate or retrieve data through digital databases.[40]

Finally, *language* is a seemingly basic and yet powerful barrier to the implementation of data journeys, particularly in a situation where scientific research is increasingly globalized. It is important to remember that most vehicles currently employed for data journeys, whether they are databases, software, Excel spreadsheets, or publications, presuppose the ability to speak English—both to take advantage of these resources and to contribute to their development. This works considerably well, given the popularity of English as a lingua franca for scientific communication; and yet it reduces access, particularly for those whose native language is very distant from West Germanic languages—such as Mandarin Chinese speakers, who, when considering all Mandarin dialects together, vastly outnumber native speak-ers of English. The language barrier becomes even more problematic when one considers the activities involved in developing and assessing metadata, without which data dissemination and reuse would not be possible. As I discussed in chapter 4, the informal knowledge about research procedures and environments that metadata need to capture is already hard to formal-ize in one's own language, which is also the language in which discussions are held, instruments are calibrated, protocols are developed and passed around, and materials are discussed and classified. Translating this knowl-edge into English, and evaluating metadata provided in that language, poses a major additional challenge to the hundreds of thousands of researchers who speak other languages in their everyday work.

One glaring consequence of these obstacles is that data dissemination practices tend to be coordinated and thus controlled by few well-resourced locations, such as the European Bioinformatics Institute in the United King-dom and Harvard in the United States, which can afford to invest in data infrastructure as a way to further their prestige and international visibility. Such unequal participation in data journeys has epistemologically significant implications. Perhaps the most important of those is that online data collections tend to be extremely partial in the data that they include and package for travel. The incorporation of data produced by poor or unfashionable labs, whether in developed or developing countries, is very low. Despite curators' best efforts, model organism databases tend to display the outputs of labs within visible and highly reputed research traditions, which deal with highly tractable data formats—which, paradoxically, some biologists view as cheap and unreliable forms of data in the first place, as discussed by Krohs under the heading of "convenience experimentation."[41] This in turn makes some types of data much more visible and widely accessible than others, thus turning them into central resources around which to structure research in the future (a clear case of existing data affecting the development of research and future notions of what counts as data, rather than the other way around). Furthermore, model organism research is itself the result of attempts to restrict the complexity of biological phenomena to mechanisms and processes observable only under highly standardized conditions on a limited number of species. As such, it excludes many key sources of information about the biological world, including data about the environment, the behavior of organisms, their evolutionary history, and the ways in which factors such as climate, population size, and ecosystems affect processes of reproduction and development. As I illustrated in the case of iPlant, model organism databases are being updated and refashioned to incorporate more and more of these additional data sources, to help foster biologists' understanding of the interactions between genes, organisms, and environments. Nevertheless, the historical circumstances under which this type of research has emerged and thrived make "omics" data into the unavoidable starting point for these attempts, and even sophisticated community databases such as iPlant are still far from constituting all-encompassing sources for the data produced by contemporary life science.

This is further underscored by the fact that these databases are not even trying to include data accumulated in previous periods of the history of biology. The exclusion of old data (sometimes called "legacy data") is perfectly understandable on practical grounds, given the difficulties involved in accessing and assembling such results, their dependence on obsolete technologies and media, the enormous variety in their formats, and the concerns surrounding their usefulness, which is often questioned given the ever-changing research contexts in which data are obtained. This compares to similar situations in other areas, most notably high-energy physics where data from particle accelerators that have been discontinued are no longer available in usable formats (e.g., the data accumulated from the predecessor to the Large Hadron Collider at CERN are kept on floppy disks and thus rarely consulted). And yet it brings databases even further from being a comprehensive source of information that encompasses scientific results not only synchronically, but also diachronically. It also raises questions about the long-term sustainability of these efforts, given the ever-increasing pace of data production in the life sciences and the difficulties hitherto experienced by curators in acquiring sufficient funding to adequately maintain and update their infrastructures, so that they adapt to shifts in propositional and embodied knowledge.

The realities of data journeys, particularly the demands that they place on scientific groups and institutions, are stark reminders of the extent to which they are embedded within resilient political, cultural, and economic structures. As a result, data practices may well be used to reinforce, rather than challenge, current power relations in science: but whether this turns out to be the case depends on how data journeys are managed and financed and who is involved in their development. This realization should dampen the hope that the availability of technologies such as the Internet and online databases can single-handedly transform the ways in which research is organized and evaluated. Data-centric biology should not be read as a case of technological determinism or "technical fix," where the emergence of new tools is the main cause for scientific, regulatory, and political shifts in how science is done, and a straightforward solution to long-standing problems.[42] Technology is certainly a key component of data journeys, and many of the features I discussed so far—most notably, the opportunity to swiftly disseminate large quantities of data and to analyze them through sophisticated classification systems such as bio-ontologies—could not have been realized without innovations in computing, software, and communications. At the same time, the conceptualization and development of such technology over the last fifty years has been deeply intertwined with scientists' shifting understandings of their fields, of data as research materials, and of the way in which science should be communicated and organized.[43] In this sense, the introduction of computers as essential technologies in biology follows the general pattern identified by Jon Agar as follows: "Computerization, using electronic stored-program computers, has only been attempted in settings where there already existed material and theoretical computational practices and technologies."[44] In molecular biology, the popularity of metaphors such as coding and information has shaped researchers' preference for digital means to express and communicate sequence data; early attempts to standardize materials and procedures involved in data production, such as those documented by Kohler in the case of *Drosophila* research in the

1920s,[45] have paved the way for the development of stock centers and metadata that are essential to current data journeys in biology; and the emergence of large collaborative networks around data production (exemplified by the Human Genome Project but anticipated by the founding of several model organism communities in the previous decades) has provided a blueprint for how research centered on "omics" data should be organized and managed. In turn, these developments favored a high level of coconstruction between computing facilities and knowledge production strategies.[46] In areas where similar organizational and conceptual shifts have not taken place, such as in some parts of theoretical biology, behavioral psychology, cell biology, immunology, physiology, and field studies, there are much larger gaps between the opportunities offered by cutting-edge digital technologies and the realities of biological data production and use.[47]

If these gaps are not acknowledged and addressed, data dissemination strategies risk acting as a magnifying glass for existing inequalities and disparities in research, rather than as means through which differences and disagreements can be voiced and scientific pluralism can be harnessed to expand the evidential value of data. Given the limited extent to which biological data are currently disseminated via databases, it seems irrational and damaging to assume that what counts as data in the first place should be defined as whatever can be fitted into highly visible databases and that results that are hard to disseminate in this way do not count as data at all, since they are not widely accessible. And yet the increasing prominence of databases as supposedly comprehensive sources of information may well lead some scientists to use them as benchmarks for what counts as data in a specific area of investigation, with worrying consequences given the amount of results that are excluded from such resources.

As discussed in chapter 2, this tendency is reinforced by wider political and economic forces, with governments and corporations powerfully drawn by the prospect of assembling centralized strategies to capture all available evidence on any given topics. The Elixir project, launched by the European Commission in 2013 to coordinate the collection and dissemination of all biological data produced through European funding, exemplifies this yearning for top-down, large-scale initiatives. If flanked by adequate efforts to develop classificatory theories and metadata, as well as reliable tools for the visualization and modeling of data, such coordination may contribute substantially to improving the scope, efficiency, and sustainability of data journeys. However, the scale of these efforts makes it exceptionally difficult to create standards that may accommodate the vast diversity of epistemic cultures involved and still remain accessible for scrutiny and feedback to such a wide variety of expertises. Whenever they select standards, labels, and data sources, and strategize over future priorities and funding bids, projects such as Elixir need to acknowledge that their decisions have the potential to determine who is included and who is excluded from contributing to data-centric biology. In the absence of extensive empirical research and explicit reflection on this phenomenon, there is a tangible danger that reliance on data-centric analysis will improve the visibility of research approaches that are already well-established, while less well-known or popular traditions get sidelined, no matter their innovative potential. In other words, by regulating the extent to which data-centric methods incorporate or exclude dissent, diversity, and creative insights, the ways in which data journeys are managed will determine the degree of conceptual conservatism characterizing this approach to biology and the extent to which it can support radically new insights.

Looking at the conditions under which data do and do not travel in biology also highlights the ever-growing significance of sampling issues in data-centric research. I already stressed that data circulated through community databases turn out to represent highly selected materials and contributions, to the exclusion of the vast majority of biological work. This selection is not the result of well-documented scientific choices, which can be taken into account when analyzing the data. Rather, it is the serendipitous result of social, political, economic, and technical factors, which determine which data get to travel in ways that are nontransparent and hard to reconstruct and assess by biologists at the receiving end. Even my brief analysis has highlighted how data journeys depend on national data donation policies (including privacy laws, in the case of biomedical data); the goodwill and resources of specific data producers; the ethos and visibility of the scientific traditions and environments in which these producers work (for instance, biologists working for private industries may not be allowed to publicly disclose their data); and the availability of well-curated databases, which in turn depends on the visibility and value placed on them (and the data types therein) by government or relevant public/private funders. Unless scientific institutions find a way to improve the accountability of processes of inclusion in biological databases, the latter will continue to provide a privileged dissemination platform for a minority of irrationally selected datasets, thus again encouraging an inherently conservative and implicitly partial platform to discovery.

This partiality speaks to the issue of bias and error in research, which in turn raises concerns about the quality and reliability of data being circulated online. As data journeys become more complex, thanks to the efforts of large numbers of individuals embedded in increasingly diverse research cultures, the margin of error involved might considerably increase. As I pointed out in chapter 4, the distributed nature of scientific reasoning underlying these forms of data dissemination presupposes high levels of trust among the individuals involved, as nobody can control or even identify the decisions made by other contributors and understand their theoretical implications. In such a system, error can creep in through explicit breaches of trust (such as the fraudulent dissemination of data generated through malfunctioning instruments and/or falsified to look different from their original form) or, more frequently, mundane mistakes that are even harder to spot (such as those made when transferring data across media, for instance typos when keying numbers on a computer or when annotating new entries in a database). Data journeys also introduce numerous biases related to the variety of methods used for data collection, storage, dissemination, and visualization. One blatant example emerging from my analysis is the tendency to consider model organism databases as role models and obligatory passage points for developing new data-centric methods of analysis. Given the theoretical and material constraints under which model organism research takes place, this tendency may well be viewed as fueling the transformation of databases into black-boxes, whose functioning is determined by assumptions that are not explicitly acknowledged and thus obscured from the users' view.

A concrete instance is provided by the incorporation of data acquired from clinical research on human subjects—which by and large document pathogenic states—into databases developed to collect data on nonpathogenic organisms, such as most model organisms.[48] As I also noted in my discussion of translational integration in plant science, decisions about how data should be handled and disseminated play a significant role in determining what organisms count as healthy or pathogenic. Unless the underlying shift in the definition of what counts as a research organism is acknowledged and critically evaluated, these decisions may come to affect how researchers understand and treat disease in ways that are not scrutinized or calibrated in relation to the specific situation in which researchers operate. This kind of black-boxing may well give rise to a "house of cards" situation, where trust is given to a system whose foundations are neither stable nor adequate to support inferences in the absence of constant checks and scrutiny. This is particularly concerning given that the standards set for data dissemination through databases are starting to affect the way in which future experimental research is planned, as researchers committed to the Open Data movement settle on experiments that will enable them to build on existing data resources and use labels provided by existing bio-ontologies and metadata.

Several advocates of the power of data-intensive methods have dismissed the epistemic significance of these issues, by pointing to the scale of such data collection as taking importance away from the singular qualities of individual data points: in other words, it does not matter if we insert a few faulty data in our repositories, because these will be made statistically insignificant by the rest of the data gathered therein.[49] As I argue in the next chapter, I agree that single data points matter increasingly little in the face of the vast collections assembled by community databases, and that it is rather the way in which data are arranged, selected, visualized, and analyzed that determines which trends and patterns emerge. However, I take issue with the assumption, underlying many such arguments, that the diversity and variability of data thus collected is enough to counter the bias and error incorporated in each of these sources. Indeed, some commentators argue that large data collections are self-correcting by virtue of their comprehensiveness, which makes it probable that incorrect or inaccurate data are rooted out of the system because of their incongruence with other data sources. In contrast to this hopeful assessment, my arguments about the inherent imbalances in the types and sources of data assembled within biology casts doubt on whether such data collections, no matter how large, are diverse enough to counter bias in their sources. If all data sources share more or less the same biases (for instance, by relying on microarrays produced with the same machines or by considering only data pertaining to a given species), there is a chance that bias will be amplified, rather than reduced, through the assemblage of such big data; and again, given the distributed nature of data handling and interpretation, it may be hard for any one individual to be able to identify such bias and critically assess its significance. This is arguably why database curators put so much care in enabling users to assess for themselves whether data are accurate and reliable and why database users attribute so much importance to being able to form their own judgment on the quality and sources of data found on the Internet.

Current claims about the potential democratizing power of data technologies, as well as the revolutionary effects that big data and data-intensive methods could have on scientific discovery, need to be measured against the concerns I have just discussed. Having access to a lot of data is not the same as having access to all of them, and cultivating such an illusion of comprehensiveness is a risky and potentially misleading strategy within biology—as many researchers whom I have interviewed over the last few years have repeatedly pointed out to me. This analysis underscores the importance of what the Royal Society called "intelligent sharing":[50] being able to access data is only fruitful to biologists if the data are appropriately packaged for travel. Because of this, data-centric approaches such the ones described here may well be limited in their applicability and should not be assumed to affect the whole of biology in the same way. Their impact is strongly felt within experimental biology, where data collection and interpretation are predominant preoccupations and the establishment of data infrastructures has been relatively well funded; indeed, model organism research has served me well as an empirical terrain to investigate data centrism in its most sophisticated forms. At the same time, my discussion of cross-species and translational integration illustrates how, even within this area, the success of data-centric approaches will eventually be determined by the efficiency with which structures originally devised to collect genomic and metabolomic data will manage to incorporate other kinds of inputs, such as environmental and behavioral data on groups and ecosystems.

Hence my analysis of data journeys, particularly of the role played by databases in facilitating them, raises questions that all participants in data-centric biology should keep clearly in mind when setting up and developing their projects. What opportunities are opened by the use of online databases, which may not have been available in their absence? At what price are such opportunities introduced—that is, what bottlenecks does the use of online databases impose on future research, especially in areas that are not yet strongly affected by the introduction of these tools and are thus not actively engaged in their development? Do these tools help biologists pose new questions, or do they instead encourage them to lose sight of the content and implications of the research being carried out, and thus of the overarching goals and creative challenges for their discipline? And to which extent are these tools backward looking, in the sense of relying on old assumptions rather than encouraging new ideas and approaches? The answer to each of these questions will depend on the specific project, data, infrastructures, goals, research situations, and institutional mechanisms involved in any one instance of data-centric biology. As I argue in the next section, it is participants' awareness of the potential pitfalls of this approach, and their willingness to explicitly question the biases and limits characterizing their projects, that will determine the long-term prospects of this research mode in the future.

6.3 The Novelty of Data Centrism:
Opportunities and Future Developments

I have hitherto portrayed data-centric science as strongly dependent on participants' ability to critically assess the embodied and propositional knowledge involved in data journeys. This runs counter the expectation that the integration of data coming from vastly different sources and research cultures can even be seamless or unproblematic. Data do not easily flow along the channels devised for their dissemination and reuse but rather undertake unpredictable journeys full of obstacles, interruptions, and setbacks, which are addressed in creative and labor-intensive ways by the many researchers involved at each stage of travel. It is these efforts to get data to travel that result in epistemologically relevant changes to scientific inquiry, including shifts in what counts as data in the first place, and for whom; what counts as relevant embodied knowledge, and with which implications for data interpretation; and what counts as theory, including the conceptual commitments undertaken to make data portable across contexts.

These insights can be generalized by defining data-centric biology as focusing more on the *processes* through which research is carried out than on its ultimate outcomes. Doubtless, biologists engage in data dissemination and mining with specific questions in mind, such as the desire to understand the relation between genes, organisms, and environment and to be able to manipulate such relations; and theoretical frameworks such as cell theory and evolutionary principles continue to play a foundational role in shaping research directions, for instance by becoming part of classificatory theories. Nevertheless, data-centric biology does not seem to be grounded on one overarching ontology of life. Rather, this approach is focused on epistemological questions about how organisms can, and should, be studied and explicitly strives to incorporate a diversity of ontological commitments, which do not have to overlap or even be compatible with each other in order to inform data interpretation. The ways in which data journeys are implemented unavoidably shape how biologists make sense of data, but this does not prevent data journeys from generating surprises and encouraging pluralistic interpretations. As we have seen, the development and use of community databases is accompanied by a desire to facilitate data journeys without overdetermining the prospective biological significance of data. The more opportunities biologists have to exercise their critical judgment on data retrieved through databases, the more likely it is that data thus disseminated will be interpreted in a variety of different ways, thus expanding their evidential value. By the same token, many database curators insist that their work does not aim to create all-encompassing frameworks for data interpretation, in which users will be forced to adopt a specific way of thinking about the world, but rather to enable exchanges and communication among epistemic cultures with diverse, and even potentially opposing, beliefs, and commitments. Indeed, community databases are often valued because they enable researchers to check findings coming from different sources against each other (a process that Alison Wylie discusses under the heading of "triangulation"), to improve the accuracy of measurement and determine which data are most reliable.[51] And while data mining does enable scientist to spot potentially significant patterns, biologists rarely consider such correlations as discoveries in themselves and rather use them as heuristics that shape the future directions of their work.[52]

When taking this perspective, the case of model organism databases can be interpreted both as a warning of the potential pitfalls of this approach and as a test case for the circumstances under which data journeys enhance biological understanding. In its most productive forms, the implementation of data journeys involves the use of computational tools to raise awareness of the conceptual, material, and institutional scaffolding required to package and interpret data, rather than hiding those aspects away. This may in turn raise the epistemic status of discussions over data handling and dissemination practices, so that they are recognized as crucial to knowledge production—and as crucially open to scrutiny by peers and the general public—in the same way as discussions over the validity of selected datasets as evidence for specific claims. Examples of this trend are the increasing attention that major scientific journals, such as *Nature* and *Science*, dedicated to cases of fraud and error in data mining and published research findings,[53] and the concerns with the experimental replicability of results that I discussed in chapter 4. We have also seen how this attitude affects how data centrism deals with synthesis, by which I mean the ability to transform large masses of unorganized and incongruous information into meaningful knowledge claims. Rather than focusing on producing overarching theoretical frameworks, data-centric biology attempts to create homogeneous standards and common infrastructures through which data can be brought together and integrated in response to specific queries, and thus in ways that may vary considerably depending on who asks the questions and why. Again, the emphasis here is on developing common procedures, rather than common theoretical frameworks or an overarching understanding of how data should be valued and interpreted.

I view the opportunity to enhance researchers' critical and reflexive attitude to data handling practices as the most significant outcome that data-centric methods may hope to achieve, and the one that may have the most interesting implications for biology. This is reflected in the *temporal*, *spatial*, and *social* orders that data-centric approaches tend to impose on biological research. The popularity of the term "data-*driven*" research is arguably due to the emphasis placed on consulting available data resources as an obligatory first step for any research project, whose outcomes are supposed to affect how researchers choose to plan future directions, which questions they address, and which additional data they produce (and how). I have illustrated how this move does not amount to straightforward inductive reasoning, therefore hardly fitting the idea that new research may be primarily or even uniquely driven by existing data; and yet, the heuristic power of data mining is undeniable and is indeed being adopted as a starting point for research planning in all the areas of biology where sufficiently powerful databases have been developed. Hence the remarkable recent shift from labs organized primarily around bench work to labs in which biologists spend a lot of their time in front of computer screens. In addition to being a key tool for recording and analyzing data, computers have become indispensable means for consulting existing datasets of relevance to prospective research projects. By reshuffling the temporal order in which computers are used in research, data-centric biology also impacts the space in which research is performed. Further, the multiple ways in which data may be de- and recontextualized mean that some loci of data travel may become as important to data travel as the original locus of data production, resulting in a variety of research spaces significantly affecting ways in which data are ultimately interpreted. Unavoidably, these temporal and spatial reconfigurations of research procedures have social repercussions, with many more researchers potentially able to participate in data journeys than ever before. Remarkable examples of this trend are projects like Synaptic Leap, an online platform that aims to engender "massively distributed collaboration" among biologists, biochemists, and medical researchers around the world in order to understand and treat tropical diseases such as malaria, schistosomiasis, toxoplasmosis, and tuberculosis, which are typically overlooked by pharmaceutical research because of the lack of profit incentives,[54] and OpenAshDieBack, a similarly successful attempt to crowdsource genomic data on ash trees and their pathogens, so as to accelerate the scientific response to forest infestation emergencies.[55] Even more interestingly, research projects organized around crowdsourcing are sometimes able to involve individuals with no professional qualifications in science, for instance by asking them to perform environmental monitoring by collecting data on given ecosystems, species, or diseases (as also exemplified by the photographs of diseased trees collected by hikers and landowners in the case of the translational integration analyzed above). These cases, sometimes referred to as promising instances of "citizen science," demonstrate the potential of data-centric biology to draw on groups, spaces, and resources well beyond the limited number of research sites and communities institutionally sanctioned as participants in scientific networks.[56]

It would be extremely interesting to assess how these shifts in biological practice are affecting the content of the knowledge being produced, and yet the implementation of new forms of data travel is too recent to be able to make any definite pronouncement on this issue. Insofar as data-centric research underscores the dynamic and iterative nature of biological inquiry, it does align well with the current emphasis placed on processual understandings of life and the attention paid to complexity and conceptual integration in areas such as epigenetics, metagenomics, systems biology, and evolutionary-developmental biology. One of the key tenets of the so-called postgenomic era is the recognition that life is defined by change at multiple scales, ranging from the evolutionary to the developmental and the microbial, and that the investigation of organisms needs to take full account of how these scales intersect at any point of their life cycle.[57] Just as the community databases developed to support data travel in model organism research have played a crucial role in fostering an interlevel understanding of these entities, the new temporal, spatial, and social orders emerging with data-centric biology may well improve the ways in which science captures such complexity, both across species and in relation to the environment.

This perspective helps to make sense of the ways in which data-centric biology may be considered to be novel. Attempts to develop adequate storage and communication technologies, institutions, and terminologies that facilitate conceptual and methodological exchanges stretch back several centuries. What strikes me as new today are (1) the scale and visibility of efforts to discuss, regulate, and systematize data practices, including data storage and mobility, and (2) the increasing congruence between aspects traditionally viewed as "internal" to research, such as the ways in which organisms are conceptualized and the procedures adopted to investigate them, and "external" aspects such as the embedding of research practices into specific social, cultural, and economic structures and trends. The capacity of data-centric research to intersect with both scientific and societal discourse and practices is a key factor in its current success. This is undoubtedly facilitated by the new opportunities for dissemination and organization offered by computing and communication technologies, which in turn explains why such an approach has not been developed—or at least, not on this scale—in previous periods of the history of biology. On the one hand, there is a marked parallel between data-centric conceptualizations of research practices as quintessentially dynamic and iterative, thus moving away from an understanding of the scientific method as a linear and context-invariant entity, and conceptualizations of life itself as essentially defined by change, complexity, and specificity.[58] On the other hand, there is also a parallel between scientific and cultural expectations of how knowledge production should operate in an increasingly globalized world, particularly of how conceptual, material, and institutional tools should be developed to facilitate exchanged between diverse groups. Indeed, the emphasis on the dissemination of material outputs of research that characterizes data centrism may itself be interpreted as a strategy to move past the tensions arising between increasingly specialized and diverse epistemic cultures. These tensions have proved particularly pernicious in twentieth-century biology, where the fragmentation of research into several subfields created multiple productive lines of inquiry but also serious obstacles to the coordination of available resources and the transfer of knowledge, with relatively little communication happening even among biologists working on the same phenomena (but from different perspectives). By focusing on the procedural aspects of knowledge production, data-centric biology draws attention away from scientific dissent over theories or approaches, which has typically proved difficult to resolve when framed in purely conceptual terms, and instead encourages a situation where disagreement can be transformed into a learning experience for all researchers involved (even if they agree to disagree). This strategy reflects broader political moves away from grand ideologies and toward a reconciliation of the need for international diplomacy and coordination with the recognition of cultural and societal diversity. Hence the use of data as a currency to broker interlevel and cross-species collaborations runs parallel to the political use of economic agreements—which also focus on the material exchange of commodities—as platforms through which intercultural dialogue may be achieved without necessarily challenging the legitimacy and authority of existing systems of governance. And just as in the political case, the scientific success of this strategy ultimately depends on how, where, and when it is implemented.

These reflections hopefully illustrate how my emphasis on the worries associated with data-centric biology does not amount to a dismissal of its innovative potential. Much of the promise of data-centric methods has yet to be realized in the case of biology, and whether this will happen depends heavily on two factors. The first is the way in which data journeys are managed and facilitated. Databases play a crucial role in this, by regulating the ways in which data are selected, retrieved, and evaluated. Depending on the care and labor involved in their packaging, data movements retain the potential to encourage a diversity of interpretations, regardless of how highly choreographed and standardized they are. The second factor is the extent to which researchers, particularly biologists reusing data found online, acknowledge and address the challenges that I have outlined in the previous section. Depending on the degree of reflexivity with which they are developed and applied, data-centric approaches can stimulate an increase in the accountability and rigor of research practices or instead be used to black-box data packaging activities and thus diminish critical participation. Hence all contributors to data-centric analysis and data journeys need to take some responsibility for the development of the relevant computational tools. Biologists in particular need to engage in understanding their functioning at the minimal level required to be able to provide feedback—a commitment that, given the considerable time and resources involved, should in turn receive significant support and recognition by scientific institutions and funding bodies.

7   Handling Data to Produce Knowledge

Several international organizations have argued that future advances in biomedical research depend on scientists' ability to consult and use as much data as possible and that data sharing on a global scale is the best way to "advance science for the public good."[1] This position resonates with the idea that the availability of big data, and the sophisticated ways in which they can be procured, integrated, and analyzed, is prompting revolutionary changes to the methods and reasoning processes involved in knowledge production. Qualifying exactly what the revolution consists of, however, proves hard. I have shown how providing access to data does not, by itself, guarantee their effective use as evidence. The availability of large quantities of data, in formats amenable to computation and statistical analysis, has certainly enhanced scientists' ability to identify patterns and correlations. At the same time, researchers wishing to interpret the scientific significance of such patterns—what they may indicate about the world and thus about the claims that data may be used as evidence for—need to determine whether data are reliable, what relation they have to specific samples and experimental setups, and whether patterns fit previously accrued knowledge about the phenomena in question. This in turn means interrogating the conditions under which data have been produced and disseminated, including the theories, methods, materials, and communities involved in all stages of data travel.

These observations make it hard to single out a particular way of reasoning as characteristic of data-centric research. It may be argued, for instance, that data centrism fits one of the "styles of scientific thinking" introduced by Albert Crombie and Ian Hacking to identify and designate differences in methodological approaches within the sciences—or even that it consists of a style of reasoning in and of itself. This is a difficult position to defend, since several such styles are implicated in the travel of biological data, and the ways in which they intersect may vary considerably depending on the research situations at hand. Within bio-ontologies, for instance, the development of classificatory theories and the articulation of metadata involve both "the experimental exploration and measurement of complex observable relations" and "the ordering of variety by comparison and taxonomy," while the construction of retrieval and visualization mechanisms in model organism databases draws on "the hypothetical construction of analogical models" as well as "the statistical analysis of regularities of populations and the calculus of probabilities" and "the historical deviation of genetic development."[2] Attempting to tie data centrism to a specific style of reasoning therefore does not seem to be a fruitful way of capturing its epistemological features.[3]

Another option is to consider whether data-centric research fits mainstream accounts of inference within the philosophy of science, such as induction, deduction, and abduction. We have already seen how appeals to inductive inference cannot do justice to the complex and distributed nature of the reasoning thus enacted, in which data certainly play a star role but could not be used to generate biological insights in the absence of the conceptual, material, and social infrastructure constructed to facilitate their handling. Neither does data centrism abide to traditional accounts of deductive reasoning, since the evidential value of data is not assigned on the basis of inference from first principles or law-like generalizations but rather depends on how data journeys unfold and on which destinations data end up reaching. Abduction, the process through which a hypothesis is chosen as the most plausible explanation for a given observation, constitutes a more promising fit. Abductive reasoning involves considering several possible explanations and choosing one on the basis of an overall assessment of the circumstances in which the observation occurred and the context in which it is being interpreted.[4] This suits the underdetermined nature of data journeys and accounts for the expectation that different users accessing the same dataset may disagree on its interpretation. However, it does not capture the interrelation of technologies, commitments, institutions, expertises, and materials affecting which observations are available to whom and which types of claims are formulated and evaluated as plausible explanations.

Associating data centrism to a specific pattern of reasoning, no matter how broad, does do justice to the complex ensembles of practices and conceptual steps involved in data journeys and their interpretation across diverse contexts. In this chapter, I therefore move away from attempts to formalize the structure of inferential activities and reflect instead on the material, institutional, and social conditions that expand and/or transform the evidential value of data and thus the diversity of reasoning processes through which they can be interpreted.[5] I contend that data centrism can be described not by reference to a specific method, technology, or reasoning pattern but rather as encompassing a particular model of attention within research, within which concerns around data handling take precedence over theoretical questions around the logical implications of given axiom or stipulation.[6] Data-centric biology focuses on diversifying the number and types of research situations in which the evidential value of data can be evaluated, in the hope of increasing the chance that different interpretations may emerge and thus that the same dataset may inspire and/or corroborate one or more discoveries.

To clarify this argument, I start with a discussion of the notion of context, which philosophers often evoke to acknowledge the material and social circumstances in which research is carried out and yet proves unhelpful to investigating the dynamic intersections between such circumstances and the conceptual work involved in extracting knowledge from data. My study of the practices involved in the circulation, integration, and interpretation of data illustrates that what counts as their context can shift considerably throughout their journeys, and the ways in which researchers perceive and manage these variations matter enormously to both the development and the results of scientific inquiry. I thus suggest abandoning the idea of context in favor of John Dewey's notion of *situation*, which emphasizes the inherent instability and evolving character of the conditions that turn out to be of relevance to any given research activity and thus constitutes a better framework to account for how data acquire and change evidential value, and for whom. In closing, I discuss how producing knowledge involves situating data in relation to elements of relevance for specific interpretive acts, which may include materials, instruments, interests, social networks, and norms for what counts as evidence—as well as specific ways of conceptualizing and valuing knowledge itself. The relevance of concerns around the organization and visualization of data, which has long been present in biology but has acquired new prominence in the digital age, constitutes a key characteristic of data centrism and a potential challenge to philosophical readings of scientific synthesis and systematization.

7.1 Problematizing Context

The most prominent philosopher to have tackled the question of what constitutes context for scientific research is arguably Hans Reichenbach. In his
1938 book *Experience and Prediction*, Reichenbach famously distinguished
the unpredictable and partly serendipitous set of processes involved in generating knowledge, which he called "context of discovery," from the post-
factum reconstruction of the reasoning that validates a given discovery, as
typically found in scientific publications, which he called "context of justification."[7] Some philosophers and historians voiced strong criticisms of
this approach, accusing Reichenbach of placing undue emphasis on textual
accounts of research and taking philosophical attention away from the study
of scientific practice.[8] Most analytic philosophers in the Anglo-American
tradition, however, adopted Reichenbach's distinction precisely because they
supported a theory-centric view of science grounded on the analysis of
justifications, to the exclusion of processes of discovery.[9] This tendency to
separate rational reconstructions from scientific practice persists today, with
many scholars assuming that the core of science consists of its conceptual
foundations, such as ontological assumptions, laws of nature, and long-
standing explanations, while material and social aspects of inquiry, such
as those documented within the historical and social studies of science, remain peripheral. "Context" thus becomes an umbrella term for all the elements of research whose significance needs to be acknowledged as affecting
knowledge production and yet is not to be the main focus of philosophical
attention—which should instead be directed to the structure and content of
the theoretical knowledge that is being created. Whether it is made explicitly
or implicitly, this hierarchical ordering of what matters in scientific research
continues to license philosophical disinterest in the relations between conceptual, material, and social practices. This is underscored by the popularity
of related distinctions such as that between internalistic and externalistic
approaches to scientific inquiry, where the latter focus on serendipitous, concrete, and informal aspects of little philosophical interest.[10]

Practice-oriented accounts of science have countered this trend chiefly
by highlighting the epistemological significance of material settings, objects,
instruments, and activities. For instance, Joseph Rouse provided one of the
first philosophical reflections on the research laboratory as "the place where
the empirical character of science is constructed through the experimenter's
local, practical know-how."[11] In a similar vein, Rheinberger used case studies from the history of biology to dissect the role of "things" in science
and persuasively argue for the need of an "epistemology of the concrete";[12]

and Chang depicted eighteenth- and nineteenth-century debates over the nature of temperature and chemical binding as illustrating the intersections between different systems of practice "formed by a coherent set of epistemic activities performed with a view to achieve certain aims," where the emphasis is on the shift from propositions to activities as the central constituents of knowledge production.[13] These views have strongly inspired my perspective and helped me particularly in framing the relation between propositional and embodied knowledge in chapter 4. Even these scholars, however, have not paid as much attention to documenting the role of social and institutional dynamics in knowledge production, as they devoted to its material and performative features.[14] This is largely a matter of emphasis, as both Rheinberger and Chang recognize the importance of social epistemology to understanding scientific inquiry;[15] and Rouse has been a vocal defender of the importance of considering material, conceptual, and social practices—and particularly politics—as crucial to the production of knowledge.[16] Beyond those instances, social and institutional circumstances are typically accorded much less philosophical attention than other aspects of scientific inquiry.

In this sense, and despite the progress made toward accounting for the actual conditions under which research is conducted, social scientists have reason to complain that philosophy of science remains for the most part detached from the world. It is rare to find studies that connect forms of reasoning and conceptual choices with specific features of the social, political, and economic environment in which they take place.[17] While an increasing number of philosophers appeal to the "context-dependence" of scientific inquiry, few question whether a distinction between content and context is even possible, and the extent to which any such separation is spatiotemporally contingent. The notion of context is stable enough as to permanently relegate some aspects of the environment to the background, as factors that have no direct bearing on the nature of knowledge claims. At the same time, it is hazy enough to satisfy worries about possible exclusions—the idea being that context can absorb any circumstance, whether or not its significance for research has been explicitly considered. When interpreted in this way, invoking the term "context" becomes a way to sweep worries about the complex and ever-changing nature of research environments under the carpet, thus separating the analysis of scientific claims and methods from concerns about who produces and handles them at any point in space and time, for which reasons, and subject to which constraints.[18]

Helen Longino has provided a detailed argument against such tendencies, in which she asks philosophers of science to explicitly challenge their well-engrained commitment to a "rational-social dichotomy" and develop "an account of scientific knowledge that is responsive to the normative uses of the term "knowledge" and to the social conditions in which scientific knowledge is produced."[19] Following in her steps, I propose an account that does justice to the critical role played by socioeconomic conditions and institutional frameworks in the development of data-centric biology, and thereby decouples the analysis of research practices from a priori judgments about what may constitute the "epistemic core" of scientific inquiry. As my investigation of data journeys illustrates, it is impossible to make sense of the epistemological implications of data handling practices without considering the conditions under which data are packaged, disseminated, and analyzed. These conditions include material and conceptual scaffolding, such as the theories used to label data, the assumptions made about their evidential value, the concrete and virtual infrastructures used to disseminate them, and the variety of research locations and instruments involved. They also include social aspects, such as the emergence of regulatory guidelines for data travel, the sanctioning of specific expertises as responsible for each stage of data journeys, and the extent to which research institutions value and incentivize data donation and reuse; economic factors, such as whether the data in question have market value as commodities, who stands to gain from their dissemination and analysis, and how; and political and legal concerns, including the use of openness to enhance transparency and debates over data ownership and the pros and cons of collecting and mining personal data. These aspects are highly dynamic and part of a continuous process of social change that is hard to capture without artificially stabilizing it.[20] Subsuming all these circumstances under the heading of context is not only simplistic and unhelpful but epistemologically suspect as the very identification of objects as data depends on these factors and, like them, is constantly susceptible to change. Accordingly, I propose moving away from the notion of context and its characterization as a stable entity that can be clearly identified in relation to specific research outputs.

7.2 From Contexts to Situations

It is interesting to note that Reichenbach himself acknowledged the dynamic nature of scientific inquiry.[21] He was particularly interested in the relation between how scientists investigate a problem and how they communicate the results of their investigation to others: in his words, "the well-known difference between a thinker's way of finding a theorem and his way of presenting it before a public."[22] Reichenbach thus recognized that what counts as scientific claim depends on the publics to which it is addressed and the ways in which those publics can be engaged and convinced of its plausibility, which in turn include social norms and institutions such as journals and learned societies. In what follows, I pursue this intuition and use it to highlight the significance of the circumstances and procedures for scientific communication in determining what is treated as data, when, and by whom. To this aim, I build on the work of John Dewey, whose book *Theory of Inquiry* appeared the same year in which Reichenbach published his argument on contexts and yet proposed a profoundly different conceptualization of research. Dewey's starting point was to view inquiry as a process, whose contents and methods shift continuously in time depending on evolving interests, aims, skills, and roles in society of the inquirers.[23] His premise is that human beings "live and act in connection with the existing environment, not in connection with isolated objects, even though a singular thing may be crucially significant in deciding how to respond to total environment"; and thus, in science as in any other type of inquiry, "there is always a field in which observation of this and that object and event occurs." Dewey focuses on how one might conceptualize such a "field," without abandoning the basic intuition that research involves a constant reassessment of the circumstances in which it takes place, and thus that what constitutes "field" may shift considerably in the course of the development of an inquiry. There are no fixed contexts in Dewey's approach and no stable distinctions between justification and discovery, as the very institutions and norms that sanction such divides are subject to change.

Rather than talking of context, Dewey prefers to label the circumstances in which research takes place at any point in time as a "situation," or "contextual whole," in which human experience and judgment are located. Dewey specifies that a situation is "whole in virtue of its immediately pervasive quality": in other words, it includes the inquiring agent(s) and all aspects of the environment that are perceived by that agent(s) at any given moment in time and are thus relevant to the practice and activities that spur the agent(s)' inquiry at that moment.[24] Agents enter the process of inquiry by identifying a problem in their environment, which they set out to solve. The problem may be conceptual, as in the case of a contradiction or a paradox, or practical, such as those emerging from encountering obstacles that prevent a given attempt to interact with the world or achieve a desired outcome. Problems give rise to a process of inquiry, which in turn stimulates the inquirer's creativity and the pursuit of novel solutions—which constitute progress in the knowledge of the world. Dewey recognizes that the ways in which inquirers perceive and frame their problem and their purpose changes during the course of inquiry. This does not prevent it from channeling the agents' actions and determines the features to which they direct their attention. In his words again: "In actual experience, there is never any such isolated singular object or event; an object or event is always a special part, phase or aspect, of an environing experienced world—a situation. The singular object stands out conspicuously because of its especially focal and crucial position at a given time in determination of some problem of use or enjoyment which the *total* complex environment presents."[25] The "total" environment is thus not all that exists but rather the specific elements that are of relevance to each stage of a process of inquiry. Furthermore, situations move from indeterminacy to determinacy, as the nature of the problem becomes increasingly clear to the inquirers, who assemble and develop some means to tackle it: "Inquiry is the controlled or directed transformation of an indeterminate situation into one that is so determinate in its constituent distinctions and relations as to convert the elements of the original situation into a unified whole."[26]

Situations are thus historical trajectories that are circumscribed and well defined, while at the same time intrinsically transformative and directional. This characterization has three features that make it peculiarly well suited to my analysis of data journeys. First, it includes conceptual, material, social, and institutional features of the environment. Dewey makes no a priori decisions about which aspects of the world should be most relevant to the reasoning and activities involved in inquiry. His account focuses strictly on the perception, goals, and experiences of inquirers, whose selection and tracking of features of the world determine what elements of their environment are relevant to their investigation. Several factors may direct their attention and determine what they view as the situation of relevance to their research—including their ways of handling and interpreting data. Some of these factors consist of background conceptual assumptions, the technological support at hand, and the types of data to which inquirers have access; others consist of the norms upheld by the institutions and communities in which inquirers work (for instance, whether or not they are willing, allowed, or encouraged to disclose data—and when, to whom, and in which form), which in turn depend on the social composition, location, and accountabilities of the groups involved, as well as on their ways of valuing and conceptualizing the processes and results of knowledge production.

Second, Dewey's characterization parallels my relational framework by stressing how the value of any dataset, and indeed its very identification as "data," depends on the ways in which it is presented and related to other data and to the situation at hand.[27] We have seen that how data are ordered, organized, and retrieved is key to how biologists interpret their evidential value at different stages of their journeys. Many database curators have emphasized to me that a fundamental part of their work involves developing several alternative ways to organize the data stored by their infrastructures, so that users would be able to personalize their searches and compare different visualizations. All model organism databases include various ways to search and retrieve their contents, and their curators devolve much of their efforts toward diversifying search parameters and linking up to other resources that may provide additional opportunities for data manipulation (a strategy often referred to as "interoperability," as it enables the integration of data stored within different databases and/or the use of searches perfected within one resource to analyze datasets stored by another). According to the curators and users that I have interviewed, the ability to order data and compare alternative data arrangements is crucial to the functioning of a database and underpins the reasoning through which the significance of the data in question is explored and further understood. Many data infrastructures have been discontinued precisely because they failed to provide such functionality, either because the technology to do it was not yet available or because insufficient strategic thinking and resources had been put into data curation.[28] Once a data repository is abandoned and the objects stored therein cannot be effectively retrieved and situated in a process of inquiry, they effectively end their life as data and become, at best, objects of curiosity—and, at worst, waste destined to oblivion.[29]

Third, Dewey's account accommodates the idea that any one research context may have flexible, dynamic boundaries, since what belongs in a situation is subject to constant change in response to the shifting goals of the inquirer(s) and the ever-evolving state of the material and social world and is thus never fixed in time or space. At the same time, it remains possible for Dewey to pinpoint exactly what boundaries a situation has at any point in time. Situations are stable enough to be singled out and studied as unique, identifiable entities. This is an important feature of Dewey's framework, because it makes situations into something more than a vague reference to whatever environment research is located in. Situations come to denote a set of circumstances that cannot be specified in the abstract and yet can be clearly identified and described in relation to specific historical trajectories. This again resonates with my account of data epistemology, in which what counts as data cannot be defined on the basis of a priori, abstract criteria but rather can be identified in relation to specific stages of data journeys, by assessing which investigators are involved, what they regard as potential evidence, and under which conditions. Borrowing Sheila Jasanoff's useful terminology, decisions on what counts as data and what counts as context may be viewed as coproduced: the ways in which researchers perceive their environment and what is relevant to their investigation shapes their choice and use of evidence, and vice versa.[30]

Dewey's account does not imply that what counts as relevant to a given situation of inquiry is an arbitrary matter (an argument famously used by Bertrand Russell to dismiss the plausibility of this view[31]). Rather, every situation consists of a historical trajectory in which specific ways of reasoning and knowing have been cultivated and established and thus "will include those things that are an intimate part of the practice and that are constitutive of it and bear on it in a meaningful fashion."[32] Within a specialized and well-funded research field such as model organism biology, many elements of inquiry have become highly socialized. In other words, they are not necessarily embodying the preferences of each individual researcher but have rather become engrained—and thus habitual and constitutive—of the practices used by the whole community of researchers engaging in a similar type of investigations. The scientific usefulness of materials such as standard organism specimens, techniques such as microarray analysis and norms such as the sharing of data through online databases have long been tested, resulting in a high degree of institutional and social trust in these research components.[33] As a result of their widespread acceptance within science, these components tend to be promoted by research groups and institutions as features that guarantee the validity of the science being conducted and thus as privileged features of any foreseeable situation within model organism research. The selection of factors that belong to a situation is thus everything but arbitrary: it is typically shaped by a long history of decisions taken by various groups of stakeholders (scientists themselves but also industry, governmental agencies, publishers, and funders) on what expertise, training, locations, instruments, and materials constitute the "best possible conditions" to carry out an investigation. Data dissemination abides by strict standards for what constitutes acceptable data types, production techniques, communication channels, and situations of reuse within different branches of biology—as exemplified by the importance of possessing the right type of embodied knowledge to be able to interpret metadata in model organism research, which I discussed in chapter 4.

It is also important to note how this account does not assume any strict demarcation between scientific and other forms of knowledge production and positions scientific research as a specific subset of the general process of human inquiry. For Dewey, there is nothing sacred or intrinsically valuable about scientific training and professionally sanctioned expertise. For sure, the research methods developed by scientists are enormously sophisticated and reliable in a variety of situations; however, their relevance and value to specific lines of inquiry depends on the circumstances, the issue at hand, and the people involved. This insight combines nicely with Reichenbach's recognition of the importance of publics in science and with my observations about the widely diverse expertises, interests, and experiences that populate the travel of data—particularly in cases such as the translational integration I discussed in the previous chapter, where data are contributed and assessed by individuals working within and outside academia, including people who do not necessarily have scientific training but are interested in the inquiry at hand and have a stake in its outcomes.

7.3 Situating Data in the Digital Age

The fact that the characteristic features of any one situation are grounded in its spatiotemporal manifestations does not mean that it is not possible to identify typologies of situations, compare them with each other, and/or conduct an abstract analysis of their role in scientific inquiry. For instance, one may view data *production* as a specific type of situation, in which researchers collect and/or generate data as relevant to exploring/developing/verifying one or more hypotheses/models/theories/claims about phenomena. In situations of data production, attention is typically focused on obtaining data in ways that are sanctioned as acceptable and reliable by peers. The selection of appropriate materials, instruments, and techniques for producing data is therefore central, with the researchers involved being more interested in the conditions under which data can be obtained, and the motivations underlying those efforts, than the conditions under which data can be mobilized and interpreted.[34]

In cases where data are obtained and analyzed by the same research group, the situation of data production will be the same as that of data *interpretation*, but as we have seen, this is not always the case in data-centric research. In cases where data are retrieved and mined by a group who had nothing to do with their production, the situations of data interpretation multiply and become distinct from data production. As we saw in chapters 4 and 5, researchers involved in interpretation thus need to decide what attitude to take toward the situation of production—whether to attempt to reproduce it or otherwise evaluate it or whether to ignore it altogether and thus trust that it is somewhat compatible with their present goals and commitments.

Another type of situation is that discussed throughout this book: data *mobilization*, where the main concern is to decontextualize data—store, classify, order, and visualize them—so as to facilitate their future recontextualization, thus making it possible for them to travel away from the situation of production and into new situations of interpretation. We have seen how mobilization can encompass a wide variety of situations, including efforts to set up and maintain data infrastructures in all fields and walks of life, and yet there are a number of common concerns and requirements. For instance, all database curators need to worry about identifying potential users and articulating a vision for what those users will wish to see and do with the database. They also confront technical questions concerning the material resources, standards, and legal and technological regimes through which data can travel; issues of data sourcing and provision; and the need to strategize over how data provenance is traced and acknowledged. These concerns set situations of data mobilization apart from situations of production and interpretation.

Each of these three types of situation focuses on a specific temporal and spatial stage of data travel and involves an appropriate set of material instruments, practices, norms, institutions, and incentives, as well as a distinctive way of assessing and conceptualizing the social, cultural, and economic value of scientific outputs. Each of them may thus involve a different conceptualization of what counts as data in the first place and what their evidential value may be. The distinction between interlevel, cross-species, and translational data integration, which I discussed in the previous chapter, provides a good instance of this: in each of those cases, different types of data are considered to be most valuable by the actors and institutions involved, and they are packaged and moved across research situations accordingly. The journeys of clinical data on human patients constitute another obvious example. Some of these data are generated via the interaction of patients with their physician and thus document situations of confidential exchange and immediate medical need, in which data are generated to provide evidence for a diagnosis and related treatment. Once these data leave the situation in which they were obtained (the doctor's office) and start traveling to national repositories, they enter a mobilization situation where their potential value is assessed according to different criteria, such as their fit with existing statistical studies of a given population. Additionally, groups tasked with the mobilization of these data have responsibilities and worries that doctors did not necessarily have in the production situation,[35] such as ethical concerns with privacy, legal concerns about potentially illegal reuses of the data to discriminate patients, and epistemic questions around their reliability and compatibility with other types of data accumulated on the same disease (e.g., from model organisms).

This analysis hopefully makes clear that stressing the unique and local qualities of situations is compatible with, and even conducive to, emphasizing their value as a unit of philosophical analysis—in the same way as stressing the relational nature of data is compatible with, and indeed conducive to, understanding their crucial role in scientific inquiry. Situations can be described, compared, abstracted, and reproduced, all while taking account of their contingent nature and their historical and geographical location. This account runs parallel to philosophical and historical work on how concepts are abstracted, generalized, and disseminated starting from descriptions of specific circumstances.[36] For instance, Rachel Ankeny has shown how "narratives of how a particular patient's course of illness, diagnosis and treatment occurred," which biomedical researchers call "cases," may become established as reference points within a field—either because their features are deemed to be typical of other situations or because they are peculiarly atypical and thus useful as contrast.[37] Descriptions of situations can thus play a useful epistemological role by serving as comparators to other situations, which can help with finding common characteristics or significant differences, or by exemplifying specific aspects of reality, such as a causal structure or set of features that researchers can then look for elsewhere (e.g., a particular biological mechanism or pathology). Notably, scholarly literature on case-based reasoning stresses that the identification and comparison of situations plays a role in all kinds of scientific inquiry and is particularly significant for research focused on understanding variable and unstable phenomena under heterogeneous and unstable conditions, as is often the case in the life or the social sciences.[38] Data journeys certainly seem to embody these characteristics, insofar as they emerge from an assemblage of specific situations of data handling, dissemination, and interpretation, each of which needs to be examined on its own merits in order to track and comprehend the sometimes highly diverse evidential value attributed to data throughout their travels. Identifying and comparing the situations characterizing different stages of data journeys makes it possible to stress both the continuities and the discontinuities between them and thus trace the genealogy of commitments that underlies any instance of data interpretation resulting in the production of knowledge claims.[39]

Throughout this book, I have defended the idea that data centrism brings new salience to aspects of scientific practice that have always been vital to successful empirical research and yet have often been overlooked by policy makers, funders, publishers, philosophers of science, and even scientists themselves, who have largely evaluated science in terms of its products (e.g., new claims about phenomena or technologies for intervention in the world) rather than in terms of the processes through which such results are eventually achieved. These include the processes involved in valuing data as a key scientific resource as well as a social and economic commodity; structuring scientific institutions and credit mechanisms so that data dissemination is supported and regulated in ways conducive to the advancement of both science and society; and, most importantly for my present discussion, enabling researchers to situate data within a variety of contexts, within which they can be analyzed and interpreted in ways that are reliable and accountable to others. Dewey's conceptualization has helped me clarify what constitutes context for data travel and interpretation, thus moving away from a priori philosophical distinctions between core and peripheral components of scientific inquiry. Rather than settling on normative frameworks determining which factors should play a constitutive role in shaping data journeys, philosophers can unravel the epistemology of data-intensive science by identifying and analyzing the variety of situations in which the significance of data is evaluated by relevant communities.

I thus conceptualize empirical inference as the process of *situating data* in relation to elements of relevance for specific interpretive acts, which may include materials, instruments, interests, social networks, and norms for what counts as evidence. The movement of objects across such processes of inquiry defines their identity as data and, at the same time, constitutes/ transforms the situations themselves. In this account, inferential processes are only partly under the control of individual scientists. Not only is data-centric reasoning distributed across sometimes large research networks, but it involves contributions by people who do not have a formal scientific training, such as civil servants involved in deciding whether to fund scientific databases, board members assessing whether or not their company should release research outputs to the public, and anybody who contributes personal or environmental data to citizen science initiatives. In fact, some of the most sophisticated databases in biology are geared to exploit—rather than repress—the widespread pluralism in venues, assumptions, and material circumstances that characterizes this type of research. For the curators and users involved, the collection and dissemination of data becomes a means to bring diverse research traditions, each with their own norms and goals, into productive contact with each other.

In model organism biology, the best databases are those that successfully support their users' attempts to situate data into a multitude of contexts and relations, which typically involves helping users to (1) identify which existing data may be relevant to their research interests (e.g., via bio-ontologies) and (2) retrieve and assess relevant information about their provenance (e.g., via metadata). These procedures, and the underlying curatorial work that make them possible, enable biologists to critically compare and integrate their current research situation with the ones in which data have been produced and disseminated, thus helping them to "know around" data as required to interpret their biological significance.[40] The identification of patterns deemed to be biologically meaningful is deeply situation-dependent, making each instance of data interpretation into a unique, spatiotemporally located act. At the same time, aspects of any given situation can be described and compared to others, thus making it possible to generalize over several situations and to determine which interpretation may best suit the goals and knowledge at hand.[41] One consequence of this view is to abandon the idea that it is possible to pick one "best" interpretation of data irrespectively of context, both when trying to create techniques and infrastructures to make data travel and when assessing the worth and quality of specific data-centric initiatives. The most interesting techniques, labels, and infrastructures developed by data curators are aimed not at facilitating one reading of data but rather at multiplying the patterns extracted from data in the course of their journeys, which in turn requires multiplying the situations within which data can be used as evidence. Indeed, funding bodies in the United Kingdom and United States tend to evaluate databases and labeling systems by measuring their success in transporting data to different research situations, rather than by establishing whether their existence facilitated one—and only one—key discovery.

The institutional, financial, cultural, and material factors involved in the decisions of the individuals that make data travel play a key role in situating data. In this sense, Dewey's terminology can be usefully related to work on situatedness within science and technology studies, particularly Donna Haraway's arguments about the need to assess the content and social significance of scientific knowledge through a detailed understanding of the institutional and material conditions through which such knowledge is produced. Haraway defines "situatedness" as the capacity to locate knowledge claims and practices, as well as the people who embody them, within what she calls "the webs of differential positioning"—the multiple ways in which humans can perceive the world that they inhabit, depending on their specific standpoint.[42] Situatedness thus captures the ways in which scientists explicitly configure their research setting as localized gateways for developing generalizations from and comparisons between datasets—a process that can vary widely depending on which stakeholders are involved and which type of knowledge is being pursued in any given research context.[43] The study of data journeys usefully exemplifies these dynamics, since it enables us to demarcate activities officially sanctioned as scientific (i.e., recognized by relevant institutions and legitimized as such) and activities that are not part of professional science, while at the same time stressing the importance of constant traffic between these realms—as in the case of translational data integration discussed in chapter 6, where the opportunity to build scientific analysis on data collected by nonscientists was crucial to knowledge production and thus an integral part of the situation in which research was carried out. Particularly in cases where data travel far and wide, crossing disciplinary and geographical boundaries, researchers involved in data interpretation are increasingly pushed to actively identify and critically debate the situations in which their research is located and the various accountabilities and biases that such situatedness generates (for instance by discussing the adequacy of the labels used for data dissemination).[44]

What, if anything, can then be said to characterize data centrism as a specific research approach? In this chapter, I have argued that the answer to this question consists not of a specific epistemology or reasoning pattern but rather a broad vision on what scientific research is and how it should be conducted. Data integration happens locally to solve specific problems, and synthesis is grounded not on a given set of principles or ways of reasoning but rather on clusters of procedures implemented by individuals with varying interests, motivations, and expertise. Data centrism can therefore be characterized by virtue of the value it places on data as a central scientific resource and commodity and the attention that it therefore bestows on the dynamics of data handling and the use of technologies, infrastructures and institutions to enhance their evidential value. The level of attention garnered by practices of data dissemination within this research mode is historically and epistemologically novel, even if the specific tools, concepts, and norms enacted in making data travel are not.

Whether or not this conclusion holds for scientific areas beyond the life sciences—or even for all areas within the life sciences, including the ones that I have not analyzed in this work—remains to be determined. In the cases I have considered, data are generated across myriads of locations and under highly diverse conditions, and data interpretation is built through recognizing and comparing the unique characteristics of situations of data production and mobilization. This may not be the case for fields where the sources of data are few and far between and most scientific work is devoted to interpretation. The production and handling of data produced by particle accelerators in high-energy physics may be viewed to exemplify such a case, insofar as data production is fully centralized, and data mobilization may not therefore appear to involve decisions around data compatibility and relevance for further uses. However, even a brief glance at the data handling practices involved in one of the best funded and most centralized physics experiments to date, the ATLAS experiment conducted by the European Organisation for Nuclear Research (CERN), challenges easy generalizations.[45] In that case, the data to be analyzed were all produced under the same circumstances; the selection of experimental data of relevance for further analysis was performed under strict guidance from theoretical assumptions and mathematical models, and in constant iteration with simulation results; and data dissemination happened through one centralized system, the Worldwide LHC Computing Grid. Prima facie, this seems very different from the case of model organism biology, where the circumstances of data production, dissemination, and use remain highly heterogeneous despite researchers' attempts to standardize materials, labels, and approaches so as to facilitate data exchange, integration, and comparison. Nevertheless, several hundred physicists and computer scientists are involved in determining which data produced by ATLAS are retained as potential evidence, how they are formatted and visualized for further analysis, and how such selection can be tested against various kinds of unwanted bias; and these processes are carried out through many labor-intensive steps at a variety of different sites, including CERN but also the dozens of labs around the globe who receive ATLAS data through the Grid. It is tempting to suggest, therefore, that data mobilization and curation in "big" particle physics also involve significant efforts, decisions, and interpretive moves that are not fully determined by an overarching theoretical framework or centralized experimental setup. It would be fascinating to check whether, upon closer scrutiny, the data selection and dissemination procedures within ATLAS turn out to be radically different from those that I analyzed, and in which respects. Indeed, questions about the procedures and assumptions involved in data mobilization should be posed to any research effort that involves inference from data, as a way to better understand the conditions under which knowledge claims are produced and validated, by whom, for whom, and why. Focusing on data should help to steer philosophers away from their traditional obsession with scientific texts and redirect their attention toward other aspects of knowledge production, including the ways in which researchers develop and articulate embodied knowledge, and order and systematize insights obtained and managed through a variety of situations.

Conclusion

This book set out to discuss the epistemology of data-centric biology, thus questioning whether this is a unique mode of scientific reasoning and practice, and how it relates to other forms of knowledge production. I have pursued these questions by examining the conditions under which data travel and are used as evidence across a variety of research situations, taking data handling practices within the last three decades of model organism biology as my main empirical focus. As I explained in chapter 1, research on model organisms is one of the most prominent and best developed arenas for the implementation of technologies and institutional regimes for the storage and mobilization of scientific data within the life sciences and thus constitutes a close approximation to what may be viewed as "ideal conditions" for data travel—including a relatively high degree of standardization of materials, instruments, and data formats; the existence of venues and communication platforms through which different research traditions can be brought together and discussed; and relatively good governmental support over several decades. Focusing on this case enabled me to examine the immediate and potential fruitfulness of data-centric research practices in one of their most sophisticated manifestations, while at the same time highlighting how, even in such an idyllic situation, the development of adequate conditions for data travel and interpretation involves confronting serious challenges and tensions. As I hope to have shown, many such challenges have not yet been overcome and may indeed never be resolved in ways that are satisfactory to all stakeholders involved as well as enduring in time and space.

Rather than being an obstacle to data centrism, or a reason to try to move away from a research culture focused on data, I presented the provisional and contested nature of data handling strategies as one of its least surprising and most fruitful characteristics. Throughout this book, I insisted that the conditions under which data can be made to travel and be used as scientific evidence are highly variable and dependent on many types of factors, ranging from the language, conceptual scaffolding, materials, and instruments used by the researchers involved to the physical infrastructure available, the political and economic value associated to data, and the extent to which scientific and other institutions support and regulate mobilization efforts. I described such conditions as parts of data journeys, in which the emergence of communication technologies such as online databases and related professions and institutions, such as biocurators and labeling centers, determine the ways in which data can be packaged for dissemination and eventual reinterpretation. I stressed that making data travel involves the implementation of processes of decontextualization (to make sure that data are extracted from their original birthplace) and recontextualization (to make it possible for researchers unfamiliar with these data to assess their evidential value and use them for their own research purposes) that are highly adapted to the specific circumstances at hand, and in fact involve prolonged debates over how research contexts can be characterized and which factors will be viewed as most relevant for data travel by different sets of scientists. I also discussed the role played by emerging professions and institutions, such as database curators and labeling centers, in realizing these forms of packaging; the various settings in which data travel is being regulated and commodified; and the resulting multiplicity in the ways in which biological data are valued, both within and beyond the world of professional science. This brought me to highlight data journeys as a key ingredient of data centrism, insofar as they aim at—and sometimes succeed in—augmenting the *evidential* value of data by expanding the range of contexts in which they can be interpreted as relevant. Furthermore, I stressed how the value of data as evidence is deeply intertwined with their economic, affective, political, and cultural value, as attributed to them by the wide network of stakeholders implicated in making data travel. In view of this, I argued against a quest for general, universally applicable solutions to the challenges arising in relation to data travel. Rather, the very recognition of such challenges, and the extent to which they are explicitly debated both by researchers on the ground and within political and economic arenas, constitute fruitful results of data centrism. The contemporary visibility of data handling practices has successfully brought attention to the processual, dynamic, and contested nature of scientific research; to the extent to which scientists themselves disagree about methodological and conceptual commitments; and to the strong interdependence between technical decisions and political, economic and cultural interpretations of what counts as evidence, research, and even knowledge. As I pointed out in chapter 2, the current popularity of data as a scientific output is at least partly due to their role as political and economic currencies and is sustained by the constant clash in interpretation over their value. Such clashes are not only unavoidable but unresolvable, as they capture seemingly incompatible and yet coexisting features of scientific data. Data are at once technical and social objects, local products and global commodities, common goods to be freely shared and strategic investments to be defended, potential evidence to be explored and meaningless clutter to be eliminated—and the tension between these conflicting and yet perfectly adequate interpretations is what keeps debates around data and their role in science so lively and indicative of the multifaceted nature of scientific and technological expertise.

In line with this broad interpretation of the phenomenon of data centrism, this book has proposed and defended a relational theory of data— what they are, how they function as evidence, and under which research circumstances—that suits the scientific and social landscapes of data travel and use within contemporary biology. In chapter 3, I defined scientific data as any material product of research activities, which is treated as potential evidence for claims about phenomena and can be circulated across a community of individuals. Within this relational approach, it is impossible to pinpoint characteristics of data in the abstract. The only relevant properties are those *attributed* to data by a given individual or group under specific conditions, which include the material features of data themselves, the circumstances of their production, their relation to existing datasets and other research components, the interests and values of the individuals involved, and the broader landscape of institutional and economic rules and expectations in which their work is embedded. This conceptualization of data supports the assumption that their evidential value is not predetermined; rather it can be interpreted in a variety of ways—which is what makes it worthwhile to disseminate them widely in the first place, while also making it impossible to settle on a dissemination strategy that works regardless of the specific circumstances of travel. In this sense, I have argued that underdetermination is the epistemological motor of data-centric research and grounds the contemporary emphasis on both "big" and "open" data. Data centrism has the potential to bring the wide variety of regimes of practice at work in biology in dialogue with each other, in ways that are not necessarily hostile or resulting in one research tradition taking over the others. Indeed, many of the packaging procedures that I analyzed in this book are geared toward harnessing the pluralism characterizing the life sciences, thus explicitly building on the difference in backgrounds, motivations, methods, and goals characterizing different biological cultures.

In biology, it is common that most data generated through research activities remain unused, no matter how well they are packaged and disseminated. This does not affect my discussion, which focuses on the fact that by virtue of their materiality and the ingenuity of the individuals involved in making them portable, data can and sometimes do travel from context to context. Scientists can and do share, exchange, and donate datasets, as well as use data generated and collected outside the realms of professional research; and many such data can be posted online and retrieved by whoever wishes to access them. This fact does not challenge the well-known philosophical contention that there are no such things as "raw" or "theory-free" data. As I stressed in chapter 4, the visualization and subsequent use of data is certainly affected by choices made during their production—decisions about how to set up experiments or observations, which instruments to use and how to calibrate them, which data formats and labels to adopt and which tools to use for collection, storage, and dissemination—which therefore have to be documented in various ways once data leave their original context. However, the evidential value of data depends not only on how they have been produced but also on how they are subsequently labeled, how users are informed about their provenance, and the variety of expertises involved in making these decisions. Indeed, I have highlighted the importance of *distributed reasoning* in data journeys and the extent to which they foster novel forms of collaboration and division of labor in the life sciences as well as raising issues of trust, control, and accountability. In chapters 4 and 5, I also emphasized the key roles played by embodied and theoretical knowledge in making data reusable across research contexts and the difficulties that this raises when attempting to automate processes of scientific discovery.

In support of this analysis, chapter 6 provided some empirical examples of how data journeys affect practices and outcomes of biological research by comparing three different cases of data integration in plant science: inter-level integration, which is closely tied to the model organism work discussed throughout the book; cross-species integration, which builds on such work and resulting data mobility to produce new knowledge on other species; and translational integration, where data are collected, mobilized, and interpreted in light of a given social agenda, rather than fundamental questions about biological ontology. These cases hopefully provided readers with a sense of the possible destinations of traveling data and the ways in which the packaging strategies discussed in previous chapters are used to generate new claims about phenomena. Looking at specific cases of data travel also helped me illustrate how data and research situations are intertwined and interdependent, an idea whose philosophical implications I analyzed in chapter 7. On the one hand, what counts as data is shaped by the expectations, expertise and goals of the situation at hand, which also determines which kind of knowledge is obtained from the use of data as evidence. On the other hand, the specific features of the data being mobilized, as well as their packaging, has a strong influence on which research strategies are pursued, by whom, and to which effect, thus effectively contributing to shape the situations of inquiry.

In bringing the above arguments together, this book proposes a general characterization of data centrism within biology, which builds on the features of specific data journeys and yet provides an abstract framework that can serve as a starting point for the analysis of other data-centric fields. I have argued that data centrism does not consist of one specific or even unique way of reasoning about data, but rather of the institutional and scientific recognition of data as a key component of scientific inquiry, whose production and widespread dissemination constitute top priorities for the scientific community. Thanks to both technological and cultural shifts over the last two decades, this recognition has reached levels never before witnessed in the history of science. Within many scientific fields, data are now viewed as a salient component of research, whose handling deserves care and transparency. Normative and practical concerns around how data should be treated and who is accountable for such activities have flooded scientific discussions over the last two decades, making attention to procedures of data production and dissemination into one of the hottest topics within contemporary science.

Data centrism thus consists of a normative vision of how scientific knowledge should be produced in order for the research process to be efficient and trustworthy. According to this vision, many technical, institutional, and political-economic features of research should be geared toward making data travel to increase their evidential value. Such features include norms for what counts as reliable and valid data, technologies to produce data and make them portable across contexts, policy guidelines stressing the importance of sharing and disseminating data, and institutions tasked with supporting and regulating data travel. This vision, and its manifestations in contemporary biology, does not call for a new, overarching epistemology of science. Initiatives that focus on data handling practices as a key component of research have often emerged in the past, both in biology and elsewhere, and their situated nature makes them impossible to associate with one specific form of reasoning or way of carrying out research. What makes the contemporary situation unique and novel are the scale and visibility acquired by efforts to discuss, regulate, and systematize procedures and aims of data dissemination and the fit between such efforts and current ways in which political regimes acknowledge and deal with diversity and inequality in the face of globalization.

From the philosophical viewpoint, studying these efforts encourages a processual view of scientific research as a highly situated, dynamic, and practical endeavor, as well as a reconceptualization of the status of data within research and their role as scientific evidence. This latter theme is one that neither philosophers nor science scholars have investigated in depth, leaving a gaping hole in science studies literature about how to understand and research the role of data in science and technology. In response to this, I hope to have demonstrated how the powerful hold of data centrism on contemporary biology constitutes an invitation to philosophers, historians, and science scholars to take data seriously as research outputs and think of them not as inert objects with intrinsic representational powers but as entities who acquire evidential value through mobilization, and may undergo significant changes as they travel. A philosophical theory of data needs to pay attention to how data are situated within specific research contexts, under which conditions, and as a result of which packaging efforts. Thinking about data, particularly the importance, development, and outcomes of data journeys within data-centric biology, fosters an understanding of research as a set of activities that are conceptually, materially, and socially grounded and inevitably intertwined with broad political, economic, and cultural trends and attitudes. At the same time, it highlights the importance of philosophical and historical analysis within science and technology studies, insofar as interrogating the functioning, nature, and roots of knowledge production in data-centric research can help to assess the complex nexus of expectations and hype associated to terms such as big data and Open Science.

Acknowledgments

This book could not have seen the light of day without the help of my husband, Michel Durinx. The joy of sharing life with him and our children, Leonardo and Luna, has been the most important inspiration for my thinking, and I cannot thank them enough for their patience and understanding, particularly given the sacrifices that research unavoidably imposes on a young family. This book is dedicated to them. I am also deeply thankful to my parents, Luca and Fany, whose love of art, nature, history, and the mysteries of classification inspired my intellectual curiosity; my brother, Andrea; zia Helen; my closest friends; and my Italian, Greek, and Belgian family for their unwavering support.

I here bring together the results of research carried out between 2004 and 2015. During this time of frequent travels and lively intellectual engagement, I was fortunate to interact with many brilliant philosophers, historians, and social and natural scientists. I cannot hope to be able to acknowledge all who helped me along the way, as much as I cannot hope to have done justice to all the insightful remarks and suggestions given to me over the years. Still, I wish to thank some of the individuals who had a direct and significant impact on the writing of this manuscript. My greatest debt is to the hundreds of biologists, bioinformaticians, and curators who discussed their work with me. Sue Rhee and her group, who generously hosted me in Stanford in August 2004, first introduced me to the complexities of data journeys; my subsequent work was marked by Sue's ability to understand the potential of computational and biological data analysis. I was lucky to continue those discussions with Midori Harris, Jane Lomax, Sean May, Andrew Millar, David Studholme, Nick Smirnoff, Klaus Mayer, Fabio Fiorani, and Michael Ashburner, among others, and to collaborate with Ruth Bastow, whose understanding of the plant science world, and its ever-changing nature, is remarkable. My second greatest debt is to Rachel Ankeny: our joint work on the history and epistemology of experimental organisms has nurtured the ideas in this book in countless ways, as has her friendship and advice over the last decade. Kaushik Sunder Rajan, Brian Rappert, Staffan Müller-Wille, and Gail Davies commented on the manuscript at several stages of writing and played a crucial role in reminding me of the multiple dimensions and implications of my analysis; the enjoyment of exchanging ideas with them has been a key motivation for this project and a significant influence on its results. James Griesemer, Maureen O'Malley, and Ken Waters provided invaluable suggestions at key moments of planning and writing and helped me articulate the philosophical concerns raised by my analysis. Hans Radder and Henk de Regt made Amsterdam a wonderful incubator for the ideas I ended up pursuing here. Mary Morgan, John Dupré, and Hasok Chang were intellectual beacons throughout this journey and shaped its development in many ways. This whole book is a belated response to Mary's invitation to investigate the "small facts" of biology. Hasok fostered my interest in integrated history and philosophy of science since my undergraduate days, and it is difficult to conceive of a more congenial intellectual environment for such an approach than the one John created at Exeter, with its commitment to pluralism, interdisciplinarity, social engagement, and a healthy dose of humor. Werner Callebaut supported my interest in model organisms and data-intensive biology from the days of my PhD, hosted me many times at the Konrad Lorenz Institute for Evolution and Cognition Research, and inspired the work on theories reported in this volume. He did not live to see the publication of this work, and his insight, kindness, wit, and friendship are deeply missed. Last but not least, my editor Karen Darling supported this project from its inception and at every step of the way, which gave me the confidence to persevere; and three anonymous reviewers provided thoughtful and constructive feedback, for which I am grateful.

At Exeter, I benefitted immensely from discussions with my colleagues and students at the Exeter Centre for the Study of the Life Sciences (particularly those participating in our Biological Interest Group, including Adam Toon, Paul Brassley, Berris Charnley, Jennifer Cuffe, Jo Donaghy, Ann-Sophie Barwich, Nadine Levin, Louise Bezuidenhout, Nick Binney, Tarquin Holmes, Gregor Halfmann, Niccolo Tempini, Susan Kelly, and Steve Hinchliffe) and from the assistance of Jim Lowe and Michel Durinx with editing the manuscript. Elsewhere, I was privileged to learn from discussions with Jon Adams, Karen Baker, Brian Balmer, Susanne Bauer, Marcel Boumans, Bill Bechtel, Anne Beaulieu, Giovanni Boniolo, Richard Burian, Ingo Brigandt, Jane Calvert, Alberto Cambrosio, Annamaria Carusi, Soraya de Chadarevian, Luciano Floridi, Roman Frigg, Emma Frow, Jean Gayon, Miguel García-Sancho, Jean-Paul Gaudillere, Elihu Gerson, Sara Green, Paul Griffiths, Lara Hueber, Phyllis Illari, Peter Keating, Lara Keuk, Ulrich Krohs, Maria Kronfeldner, Javier Leuzan, Alan Love, Ilana Loewy, Erika Masnerus, James Overton, Ted Porter, Barbara Prainsack, Edmund Ramsden, Hans-Jörg Rheinberger, Thomas Reydon, Federica Russo, Edna Suarez Diaz, Karola Stotz, Beckett Sterner, Bruno Strasser, Sharon Traweek, Orlin Vakarelov, Simona Valeriani, Niki Vermeulen, Erik Weber, Bill Wimsatt, Sally Wyatt, and Alison Wylie. Another real privilege was participating in the Global Young Academy, the Society for the Philosophy of Science in Practice, the "How Well Do Facts Travel" project and the Knowledge/Value workshop series—there is no easy way to acknowledge how deeply discussions within these four venues, and the opportunities they provided to expand my cultural and geographical horizons, have shaped my thinking. I also received invaluable feedback from audiences in Exeter, Berlin, Minneapolis, Vienna, Bristol, London, Sydney, Beijing, Bangalore, Johannesburg, Montreal, Vancouver, Pittsburgh, UC Davis, Chicago, Amsterdam, Muenster, Munich, Bielefeld, Dusseldorf, Brussels, Copenhagen, Aarhus, Edinburgh, Madrid, Rotterdam, Milan, Oxford, Cambridge, Ghent, Lausanne, Geneva, Bern, Manchester, Durham, Toronto, Paris, the University of Maryland, Copenhagen, Bologna, Milan, Leiden, Philadelphia, Sterling, the British Academy, and the Royal Society.

In terms of my subsistence and research time during this period, I am very grateful to the funding agencies and institutions that supported this work as it was progressing: the Leverhulme Trust through the projects "How Well Do Facts Travel?" (2006–8, grant award F/07004/Z) and "Beyond the Digital Divide" (2014–15, grant award RPG-2013-153); the Economic and Social Research Council (ESRC) through the ESRC Centre for Genomics and Society (2008–12) and Cross-Linking Grant (number ES/F028180/1; 2013–14); the British Academy through Small Grant SG 54237 (2009–10); the Konrad Lorenz Institute for Evolution and Cognition Research, which sponsored the conference "Making Sense of Data" in Exeter in 2011; and the European Research Council (ERC) through ERC grant agreement number 335925 (2014–19). I am also indebted to the "Sciences of the Archive" project at the Max Planck Institute for the History of Science in Berlin for financing the sabbatical that enabled me to draft the first chapters of this book. The staff and visitors of Department II, and especially Lorraine Daston, Elena Aronova, David Sepkoski, and Christine van Oertzen, generously provided me with the kind hospitality, challenging questions, and rich counterexamples that I needed in that intensive writing phase.

Early versions of some of the material in this book have been published as follows and are included with permission of the publishers:

> Chapter 1: Leonelli, Sabina. "Packaging Data for Re-Use: Databases in Model Organism Biology." In How Well Do Facts Travel? The Dissemination of *Reliable Knowledge*, edited by P. Howlett and M. S. Morgan, 325–48. Cambridge: Cambridge University Press, 2010.
> Chapter 2: Leonelli, Sabina. "Centralising Labels to Distribute Data: The Regulatory Role of Genomic Consortia." In *The Handbook for Genetics and Society: Mapping the New Genomic Era*, edited by P. Atkinson, P. Glasner, and M. Lock, 469–85. London: Routledge, 2009; Leonelli, Sabina. "Why the Current Insistence on Open Access to Scientific Data? Big Data, Knowledge Production and the Political Economy of Contemporary Biology." *Bulletin of Science, Technology and Society* 33(1/2) (2013): 6–11.
> Chapter 3: Leonelli, Sabina. "On the Locality of Data and Claims About Phenomena." *Philosophy of Science* 76(5) (2009): 737–49; Leonelli, Sabina. "What Counts as Scientific Data? A Relational Framework." *Philosophy of Science* 82 (2015): 1–12.
> Chapter 4: Leonelli, Sabina. "Data Interpretation in the Digital Age." *Perspectives on Science* 22(3) (2014): 397–417.
> Chapter 5: Leonelli, Sabina. "Classificatory Theory in Data-Intensive Science: The Case of Open Biomedical Ontologies." *International Studies in the Philosophy of Science* 26(1) (2012): 47–65; Leonelli, Sabina. "Classificatory Theory in Biology." *Biological Theory* 7(4) (2013): 338–45.
> Chapter 6: Leonelli, Sabina. "Integrating Data to Acquire New Knowledge: Three Modes of Integration in Plant Science." *Studies in the History and Philosophy of the Biological and Biomedical Sciences: Part C* 4(4) (2013): 503–14.

Notes

INTRODUCTION

1. On "big data," or the "data deluge," see Scudellari, "Data Deluge," Mayer-Schönberger and Cukier, *Big Data*, Kitchin, *The Data Revolution*.

2. "If it becomes cheaper to just collect all data required than to run after a hundred consecutive, plausible, but wrong hypotheses, starting with a hypothesis becomes an economic futility" (van Ommen, "Popper Revisited," 1). On data-driven science, see Hey, Tansley, and Tolle, *The Fourth Paradigm*; and Stevens, *Life Out of Sequence*. On the "end of theory," see Anderson, "The End of Theory."

3. Krohs, "Convenience Experimentation."

4. Bowker, *Memory Practices in the Sciences*.

5. Lawrence, "Data."

6. See special issue of *Nature* on "The Future of Publishing": 495, no. 7442 (2013).

7. For an analysis of the role of lists, see Müller-Wille and Charmantier, "Lists as Research Technologies"; and Müller-Wille and Delbourgo, "LISTMANIA." On the role of archives, see Blair, *Too Much to Know*; and Müller-Wille and Charmantier, "Natural History and Information Overload." On taxonomies, see for instance Johnson, *Ordering Life*. On museum exhibits and collections, see Daston, "Type Specimens and Scientific Memory"; Endersby, *Imperial Nature*; Star and Griesemer, "Institutional Ecology, 'Translations' and Boundary Objects"; Strasser, "Laboratories, Museums, and the Comparative Perspective." On statistical and mathematical models, see Foucault, *The Birth of the Clinic*; Kingsland, *Modeling Nature*; Bauer, "Mining Data, Gathering

Variables, and Recombining Information." For newsletters, see Leonelli and Ankeny, "Re-Thinking organisms," and Kelty, "This Is Not an Article." On databases, see Hilgartner, "Biomolecular Databases"; Lenoir, "Shaping Biomedicine as an Information Science"; Hine, *Systematics as Cyberscience*; Strasser, "GenBank"; November, *Biomedical Computing*; Chow-White and García-Sancho, "Bidirectional Shaping and Spaces of Convergence"; García-Sancho, *Biology, Computing, and the History of Molecular Sequencing*; Stevens, *Life Out of Sequence*. An overview of data gathering and processing practices across the history of several scientific fields is provided in the "Historicizing Big Data" special issue of *Osiris*, edited by David Sepkoski, Elena Aronova, and Christine van Oertzen.

8. See the discussion of the prospects for a general philosophy of science by Hans Radder, *The Material Realization of Science*, as well as the arguments for the role of case studies in philosophy by Richard Burian, "More than a Marriage of Convenience" and "The Dilemma of Case Studies Resolved," and by Jutta Schickore, "Studying Justificatory Practice," and Hasok Chang's arguments for epistemic iterativity between concrete and abstract studies of science (*Is Water H₂O?*).

9. Another label for this approach is *philosophy of science in practice*, as discussed in Ankeny et al., "Introduction: Philosophy of Science in Practice"; and Boumans and Leonelli, "Introduction: On the Philosophy of Science in Practice."

10. My contributions to these scientific debates are evidenced by coauthored papers with database developers and related experts in *EMBO Reports* (Bastow and Leonelli, "Sustainable Digital Infrastructure"), *BMC Bioinformatics* (Leonelli et al., "How the Gene Ontology Evolves"), *New Phytologist* (Leonelli et al., "Under One Leaf") and *Journal for Experimental Botany* (Leonelli et al., "Making Open Data Work for Plant Scientists"). More information about activities since 2013 is available on the Exeter Data Studies website, http://www.datastudies.eu.

11. Some of these interactions, which include Science International, the Global Young Academy, the European Research Council, CODATA, the Royal Society, the National Science Foundation, Research Councils UK, Elixir, and the UK Cabinet Office, are documented on the website of the Exeter Data Studies group, http://www.datastudies.eu.

12. Dewey, "The Need for a Recovery in Philosophy," 138.

13. See Clifford Geertz's 1973 characterization, which famously pointed to philosopher Gilbert Ryle as a crucial source of inspiration.

CHAPTER ONE

1. The workshop was organized by the Genomic Arabidopsis Resource Network (GARNet), an organization funded by the UK research councils to coordinate access to training and resources needed by UK plant scientists to keep up with the latest technological developments in data production, dissemination, and analysis. I will discuss GARNet in chapter 2. Also, note that in September 2015, shortly before this book went to press, the iPlant Collaborative announced that it was changing its name to CyVerse to reflect a broader remit to develop data infrastructures for science as a whole. For consistency and historical accuracy, I will continue to refer to it as iPlant or iPlant Collaborative in what follows.

2. See Penders, Horstman, and Vos, "Walking the Line Between Lab and Computation"; and Stevens *Life Out of Sequence.*

3. Goff et al., "The iPlant Collaborative."

4. I am not interested in evaluating the quality and effectiveness of packaging efforts, but rather my goal is to highlight the dilemmas encountered in this process and assess how theoretical commitments and material constrains shape the goals, methods, and outcomes of data dissemination.

5. I will purposefully pay little attention to the software (models, coding, and statistical algorithms) and hardware (computing machines, servers, and digital cloud services) involved in implementing labeling systems for data. These elements are also central to the packaging of data, but a detailed analysis of their epistemic role requires a book in itself and has partly been analyzed already by Mackenzie, "Bringing Sequences to Life"; Suárez-Díaz and Anaya-Muñoz, "History, Objectivity, and the Construction of Molecular Phylogenies"; García-Sancho, *Biology, Computing, and the History of Molecular Sequencing*; November, *Biomedical Computing*; and Stevens, *Life Out of Sequence.*

6. Biologists use many different terms to refer to online data infrastructures, including "database," "repository," "platform," and "grid." There are interesting nuances in the degree of complexity and organizational structure suggested by these terms, which have been discussed by several commentators (e.g., Hine, "Databases as Scientific Instruments and Their Role in the Ordering of Scientific Work"; Borgman, *Scholarship in the Digital Age*; Royal Society, "Science as an Open Enterprise"; Stevens, *Life Out of Sequence*). Within this study, I am not interested in marking essential differences between these terms. My empirical focus is on sophisticated, evolving data infrastructures, where much effort has been devoted to assembling, integrating, and displaying data from a variety of sources. The digital infrastructures used to host model organism data are most typically referred to by practitioners as "online databases," and I will abide by that terminology.

7. See Keller, *The Century of the Gene*; Kay, *The Molecular Vision of Life* and *Who Wrote the Book of Life?*; de Chadarevian, *Designs for Life*; Suárez-Díaz, "The Rhetoric of Informational Molecules"; Moody, *Digital Code of Life*; and García-Sancho, *Biology, Computing, and the History of Molecular Sequencing*, among others.

8. For examples of these critiques, see Tauber and Sarkar, "The Human Genome Project"; Dupré, *Processes of Life*; Griffiths and Stotz, *Genetics and Philosophy.*

9. This process has been thoroughly documented by Stephen Hilgartner in "Constituting Large-Scale Biology" and *Reordering Life.*

10. See Morange, *A History of Molecular Biology*; Oyama, *The Ontogeny of Information*; Barnes and Dupré, *Genomes and What to Make of Them*; Müller-Wille and Rheinberger, *A Cultural History of Heredity.*

11. See Cook-Deegan, *The Gene Wars*; Hilgartner "Biomolecular Databases" and "Constituting Large-Scale Biology"; Stevens, *Life Out of Sequence.*

12. See García-Sancho, *Biology, Computing, and the History of Molecular Sequencing*; Stevens, *Life Out of Sequence*; Hilgartner, "Constituting Large-Scale Biology." I should make clear that I do not wish to uncritically back the idea of sequencing as a science in itself. Whether sequencing projects are something more

than "mere" technical and organizational tools is still hotly debated, particularly given more recent advances in sequencing technologies. Those arguments continue to shape current discussions around the feasibility and scientific status of data-centric research, which I will discuss in chapter 6. I also do not want to argue that the prominence of data within biology is wholly due to the success of sequencing projects. There are many other reasons for the increasing visibility of data practices in all the sciences, including the availability of computing technologies as platforms for data analysis and communication (which has been examined by November, *Biomedical Computing*; García-Sancho, *Biology, Computing, and the History of Molecular Sequencing*; and Stevens, *Life Out of Sequence*) and the appropriation of data as global commodities by national governments and market structures (which I will discuss in chapter 2). What I want to highlight, rather, is the coming into being of a lively debate on these matters. Biologists have started to discuss the value of data collection and dissemination as scientific contributions in their own right. This in itself is a notable historical development, which is at least in part due to the technical and organizational resources generated to make sequencing projects possible.

13. On the scale and regimes of sequencing, see Hilgartner, "Constituting Large-Scale Biology"; and Davies, Frow, and Leonelli, "Bigger, Faster, Better?"

14. For a general characterization of big data, see Kitchin, *The Data Revolution*.

15. Seminal philosophical and historical studies of model organism research include Burian, "How the Choice of Experimental Organism Matters"; de Chadarevian, "Of Worms and Programmes"; Clarke and Fujimura, *The Right Tools for the Job*; Kohler, *Lords of the Fly*; Ankeny, "The Natural History of *Caenorhabditis elegans* Research"; Todes, *Pavlov's Physiology Factory*; Logan, "Before There Were Standards" and "The Legacy of Adolf Meyer Comparative Approach"; Rader, *Making Mice*; Weber, *Philosophy of Experimental Biology*; Rheinberger, *An Epistemology of the Concrete*; Kirk, "A Brave New Animal for a Brave New World" and "Standardization through Mechanization"; Ramsden, "Model Organisms and Model Environments"; Friese and Clarke, "Transposing Bodies of Knowledge and Technique"; Nelson, "Modeling Mouse, Human and Discipline"; Huber and Keuck, "Mutant Mice." I have provided my own characterization of model organism research in joint work with Rachel Ankeny, including Ankeny and Leonelli, "What's So Special about Model Organisms?"; Leonelli and Ankeny, "Re-Thinking organisms," "What Makes a Model Organism?" and "Repertoires."

16. Clarke and Fujimura, *The Right Tools for the Job*.

17. Kohler, *Lords of the Fly*; Ankeny, "The Natural History of *Caenorhabditis elegans* Research"; Leonelli, "Weed for Thought" and "Growing Weed, Producing Knowledge"; Rader, *Making Mice*; Endersby, *A Guinea Pig's History of Biology*.

18. Ankeny and Leonelli, "What's So Special about Model Organisms?" and "Valuing Data in Postgenomic Biology"; Leonelli and Ankeny, "What Makes a Model Organism?"

19. See, for example Bolker, "Model Systems in Developmental Biology" and "Model Organisms."

20. On boundary objects, see Star and Griesemer, "Institutional Ecology, 'Translations' and Boundary Objects."

21. The discussions surrounding the development and implementation of the Bermuda Rules for the public release of sequencing data have been documented by Harvey and McMeekin, *Public or Private Economics of Knowledge?*; Cook-Deegan, "The Science Commons in Health Research"; and Strasser, "The Experimenter's Museum." Maxson, Cook-Deegan, and Ankeny, *The Bermuda Triangle*, focuses on the role played by model organism biologists in fostering data sharing.

22. These early attempts, particularly the development of the *Arabidopsis thaliana* Database (AtDB) and the *C. elegans* Database (ACeDB), are documented by García-Sancho, *Biology, Computing, and the History of Molecular Sequencing*; Ankeny, "The Natural History of *Caenorhabditis elegans* Research"; Leonelli, "Growing Weed, Producing Knowledge"; Leonelli and Ankeny, "Re-Thinking organisms."

23. See the website of the National Human Genome Research Institute, listing the main community databases funded by the National Institutes of Health ("Model Organism Databases Supported by the National Human Genome Research Institute").

24. See Leonelli and Ankeny, "Re-Thinking organisms"; Leonelli, "When Humans are the Exception."

25. Rhee and Crosby, "Biological Databases for Plant Research"; Bult, "From Information to Understanding."

26. As I discuss in chapter 6, the remit of model organism databases has expanded considerably in the 2010s, including collaborations with infrastructures such as iPlant, which is attempting to reintroduce and integrate environmental with genomic data. It may well be that model organism biology is progressively being superseded by cross-species approaches, an event that could be regarded as further evidence for its scientific success.

27. Rhee, Dickerson, and Xu, "Bioinformatics and Its Applications in Plant Biology," 352.

28. Rhee, "Carpe Diem."

29. Leonelli, "Weed for Thought" and "Growing Weed, Producing Knowledge"; Koornneef and Meinke, "The Development of *Arabidopsis* as a Model Plant"; Somerville and Koornneef, "A Fortunate Choice"; and Jonkers, "Models and Orphans."

30. For scientific details on TAIR and its components, see, for example, Huala et al., "The *Arabidopsis* Information Resource (TAIR)"; Garcia-Hernandez et al., "TAIR"; Rhee et al., "The *Arabidopsis* Information Resource (TAIR)"; Mueller, Zhang, and Rhee, "AraCyc."

31. Rosenthal and Ashburner, "Taking Stock of Our Models."

32. Ledford, "Molecular Biology Gets Wikified"; Bastow et al., "An International Bioinformatics Infrastructure to Underpin the Arabidopsis Community."

33. TAIR and Araport continue to work together, although TAIR switched to a subscription-based service in 2013, so as to ensure its own survival. Since the development of Araport is in its starting phase at the time of writing, and its role in the *Arabidopsis* community is yet to become established, I will base my discussion solely on TAIR for the purposes of this book.

34. I am aware that by choosing to focus on model organism databases, I am overlooking parallel efforts to collect and disseminate data within the evolutionary

and environmental life sciences. Studies of such initiatives are extremely important, particularly since they capture aspects of biological reasoning that are typically excluded in the study of model organisms, such as the problems involved in collecting and interpreting data on environmental variability and long-term evolutionary dynamics (Shavit and Griesemer, "Transforming Objects into Data"; Sepkoski, *Rereading the Fossil Record*; Aronova, "Environmental Monitoring in the Making"). I discuss the implications of this choice and the extent to which data packaging practices implemented within model organism biology can be viewed as representative for biology as a whole in chapter 6.

35. The de facto pluralism characterizing biology has been widely discussed in the social and philosophical studies of science (e.g., Mitchell, *Biological Complexity and Integrative Pluralism*; Knorr Cetina, *Epistemic Cultures*; Longino, *The Fate of Knowledge*).

36. The use of the term curator is an actor's category, as most scientists whom I have worked with, and who are involved in developing databases, call themselves that. I will not delve here into the historical roots of this phenomenon nor on the relation between the title of curator and other titles used by data scientists in other areas (such as "data manager," "data archivist," "data engineer," and "data administrator"—which could be argued point to different configurations of expertise of relevance to data handling, e.g., administrative or IT skills). These are both fascinating topics deserving of further research and much more space than I can allocate to them within this book; and an excellent analysis of the IT expertise brought in by "information engineers" is provided in Chapter 4 of García-Sancho's *Biology, Computing, and the History of Molecular Sequencing*, which focuses on the making of the Nucleotide Sequence Data Library of the European Molecular Biology Laboratory, while November discusses in *Biomedical Computing* the role of postwar operations research in shaping the introduction of biomedical computing at the National Institutes of Health in the United States. I will also not dwell on the relation between curators of databases and curators of museums other than to say that while both professions involve caring for the preservation and dissemination of given sets of objects, the different nature of those objects—and the different aims for which they are stored, classified, and circulated—can determine striking differences between the practices involved in the two cases. There are situations where these two types of curation merge to some extent, as documented by Shavit and Griesemer's work on specimen preservation for biodiversity collections ("Transforming Objects into Data").

37. For a detailed study of the issues involved in the standardization of data formats, see Rogers and Cambrosio, "Making a New Technology Work," on the case of microarray data.

38. Many scientific reviews of the features of model organism databases that were published throughout the 2000s point to the epistemic diversity of user communities as a crucial factor that needs to be incorporated into database development (e.g., Ashburner et al., "Gene Ontology"; Huala et al., "The *Arabidopsis* Information Resource (TAIR)"; The FlyBase Consortium, "The FlyBase Database of the Drosophila Genome Projects and Community Literature"; Harris et al., "WormBase"; Sprague et al., "The Zebrafish Information Network"; Chisholm et al., "dictyBase").

39. Huang et al., "*Arabidopsis* VILLIN1 Generates Actin Filament Cables That Are Resistant to Depolymerization."

40. I restrict the present discussion to the bio-ontologies listed in the Open Biomedical Ontologies consortium (Smith et al., "The OBO Foundry"). I will analyze this type of label and its epistemological role in more detail in chapter 5.

41. Augen, *Bioinformatics in the Post-Genomic Era*, 64.

42. For example, nucleus is defined as "a membrane-bounded organelle of eukaryotic cells in which chromosomes are housed and replicated. In most cells, the nucleus contains all of the cell's chromosomes except the organellar chromosomes, and is the site of RNA synthesis and processing. In some species, or in specialized cell types, RNA metabolism or DNA replication may be absent" (Gene Ontology Consortium, "Minutes of the Gene Ontology Content Meeting [2004]").

43. Maddox, *Rosalind Franklin*. Thanks to an anonymous reviewer for pushing me to think harder about this case, which deserves a much lengthier treatment.

44. Morgan, "Travelling Facts."

45. See Bowker, *Memory Practices in the Sciences*; Leonelli, "Global Data for Local Science."

46. "One of the strengths of the GO development paradigm is that development of the GO has been a task performed by biologist-curators who are experts in understanding specific experimental systems: as a result, the GO is continually being updated in response to new information" (Hill et al., "Gene Ontology Annotations").

47. Goble and Wroe, "The Montagues and the Capulets"; García-Sancho, *Biology, Computing, and the History of Molecular Sequencing*; November, *Biomedical Computing*; Stevens, *Life Out of Sequence*; Lewis and Bartlett, "Inscribing a Discipline."

48. Edwards et al., "Science Friction."

49. Howe et al., "Big Data"; Rhee et al., "Use and Misuse of the Gene Ontology (GO) Annotations."

50. The workshop, which I attended, took place at the Department of Plant Systems Biology, University of Gent, Belgium (Gent, May 21–23, 2007).

51. I discuss in chapter 3 how one can conceptualize data so as to acknowledge this constant need for transformation while also fostering the trustworthiness and reliability of data for prospective users.

52. Documented by Müller-Wille and Charmantier, "Natural History and Information Overload"; and Kristin Johnson, *Ordering Life*, among others.

53. Kohler, *Lords of the Fly*; Kelty, "This Is Not an Article"; Leonelli and Ankeny, "Re-Thinking organisms."

54. Latour, "Circulating Reference"; and Morgan, "Travelling Facts."

55. This again constitutes an interesting difference between current efforts to make data travel and previous scientific attempts to mobilize data, as for instance in the case of nineteenth-century astronomy, where worries about the long-term conservation of data collections were central to their development (see Daston and Lunbeck, *Histories of Scientific Observation*). This is not to say that funders and participants in contemporary data dissemination practices are not concerned with the long-term subsistence and implications of their effort; such considerations are however hard to translate into practical strategies, given the speed of technological

change as well as the short-term nature of current funding regimes. In both Europe and North America, for instance, databases are funded in three-to-five-year cycles, which means that long-term strategies need to be staggered to respond to changing financial, cultural, and scientific landscapes. These conditions make the work of oversight organizations such as Elixir, funded by the European Commission to provide general coordination and strategic support to biological data infrastructures in Europe, as significant as it is difficult: they have the task to develop long-term visions and plans for data handling yet their operations and accountabilities are unavoidably tied to short-term funding cycles (Elixir, "Elixir").

56. Hallam Stevens makes a parallel point about the notion of "pipeline" in the production of sequence data, arguing that "the production of 'fluid' and 'universal' bioinformatic objects depends on a highly situated process through which this 'fluidity' and 'universality' is constructed; bioinformation can travel and flow only because its movement through the pipeline serves to obscure its solidity and situatedness" (*Life Out of Sequence*, 117).

57. Prainsack, *Personalization from Below*.

58. For example Livingstone, *Putting Science in Its Place*; Secord, "Knowledge in Transit"; and Subrahmanyam, *Explorations in Connected History*.

59. See for instance Raj, *Relocating Modern Science*; Schaffer et al., *The Brokered World*; and Sivasundaram, "Sciences and the Global." Such scholarship seeks to understand the category of the "global" in relation not just to differentiations of power, language, and resources but also to diverse understandings of knowledge and politics and their implications for placements of data in relation to the epistemic. I am grateful to Kaushik Sunder Rajan and an anonymous reviewer for pushing me to consider more carefully the significance of this approach.

CHAPTER TWO

1. Seminal sociological work on data infrastructures includes Hilgartner, "Biomolecular Databases"; Star and Ruhleder, "Steps Toward an Ecology of Infrastructure"; Bowker, "Biodiversity Datadiversity"; Hine, "Databases as Scientific Instruments and Their Role in the Ordering of Scientific Work"; and Edwards, *A Vast Machine*. For detailed studies of the political, cultural, and economic embedding of biological databases, see Martin, "Genetic Governance"; Bowker, *Memory Practices in the Sciences*; Harvey and McMeekin, *Public or Private Economics of Knowledge?*; Borgman, "The Conundrum of Sharing Research Data"; and Hilgartner, "Constituting Large-Scale Biology"; as well as ongoing work by Bruno Strasser, Mike Fortun, Javier Leuzan, Barbara Prainsack, Chris Kelty, Jenny Reardon, and Jean-Paul Gaudillière, among others.

2. Fortun, *Promising Genomics* and "The Care of the Data."

3. On the Celera dispute and Bermuda Rules, see Maxson, Cook-Deegan, and Ankeny, *The Bermuda Triangle*; Powledge, "Changing the Rules?"; Resnik, *Owning the Genome*; and Wellcome Trust, "Sharing Data from Large-Scale Biological Research Projects."

4. In 2013 and 2014, I along with colleagues at Exeter and Edinburgh carried out a series of interviews with prominent biologists and bioinformaticians in the United Kingdom, which clearly showed these concerns and particularly their

perception that funding bodies, universities, and commercial research partners had different, and possibly conflicting, views on who has intellectual property of data and what are the consequences of data dissemination (Levin et al., "How Do Scientists Understand Openness?"). Other studies on conflicting perceptions of data sharing include Wouters and Reddy, "Big Science Data Policies"; Piwowar, "Who Shares? Who Doesn't?"; Borgman, "The Conundrum of Sharing Research Data"; and Fecher, Frisieke, and Hebing, "What Drives Academic Data Sharing?"

5. Indeed, one of the main reasons why data plagiarism is regarded as a problem is that it makes it difficult to accurately reconstruct data provenance, which in turn may hamper future efforts to interpret data and transform them into knowledge. I have discussed the need for incentives in Open Science policies in Leonelli, Spichtiger, and Prainsack, "Sticks AND Carrots."

6. Rachel Ankeny and I discuss these issues in some detail in Ankeny and Leonelli, "Valuing Data in Postgenomic Biology." See also the analysis of the situation in the 1990s in Katherine McCain, "Mandating Sharing."

7. For an analysis of the role of technology in scientific governance, see Andrew Barry, *Political Machines*.

8. Lewis, "Gene Ontology." I analyze this in more detail in Leonelli, "Documenting the Emergence of Bio-Ontologies."

9. Bada et al., "A Short Study on the Success of the Gene Ontology."

10. Gene Ontology Consortium, "The Gene Ontology Consortium: Going Forward."

11. Smith et al., "The OBO Foundry," 1252.

12. Ibid., 1253.

13. This is particularly relevant to the OBO Foundry, a subset of OBO ontologies whose curators are actively engaged in testing and developing further rules for ontology development (Smith et al., "The OBO Foundry").

14. Lewis describes effective collaboration as an "unforeseen outcome" yet points to it as "the single largest impact and achievement of the Gene Ontology consortium to date" ("Gene Ontology," 103.3).

15. See the International Society for Biocuration website, http://www.bio curator.org.

16. Howe et al., "Big Data."

17. The first such collaboration was initiated in 2008 with *Plant Physiology*, the foremost journal in plant science, through the mediation of TAIR (Ort and Grennan, "*Plant Physiology* and TAIR Partnership").

18. In "Scientific Organisations as Social Movements," I have discussed the strategies used by members of consortia to engender collective action in support of scientific agendas.

19. I comment on the inequalities and exclusionary factors underlying these discussions in the second part of this chapter.

20. See Lee, Dourish, and Mark, "The Human Infrastructure of Cyberinfrastructure"; and Stein, "Towards a Cyberinfrastructure for the Biological Sciences." This shift has also been discussed within literature on "collaboratories" (intended as "laboratories without walls"), exemplified by Bafoutsou and Mentzas, "Review and Functional Classification of Collaborative Systems"; and Finholt, "Collaboratories."

21. Cambrosio et al., "Regulatory Objectivity and the Generation and Management of Evidence in Medicine," 193.

22. Smith et al., "The OBO Foundry," 1254.

23. Lewis, "Gene Ontology," 103.2.

24. The Research Data Alliance is emerging at the time of writing as a promising international effort in this direction. In Leonelli, "Scientific Organisations as Social Movements" and "Epistemische Diversität im Zeitalter von Big Data," I explore further the idea of consortia as filling a regulatory gap in science.

25. See, for example, Bowker and Star, *Sorting Things Out*.

26. See, for example, Ledford, "Molecular Biology Gets Wikified."

27. A rare example of a database successfully harnessing data annotations from its users is PomBase, the database dedicated to data on fission yeast *Schizosaccharomyces pombe* (http://www.pombase.org). This method of crowdsourcing works principally thanks to excellent ongoing communication between the long-term lead of the database, Valerie Wood, and the relatively small research community working on fission yeast, as well as heavy reliance on bio-ontologies, including the Gene Ontology, whose use in PomBase is supervised by Midori Harris, who previously spent almost a decade as a GO curator and is thus very well positioned to interface between GO standards and the specific requirements of yeast biologists. Reproducing such user-curator interface in bigger and more fragmented research communities presents considerable challenges.

28. Ure et al., "Aligning Technical and Human Infrastructures in the Semantic Web," 9.

29. Jasanoff, "Making Order."

30. Hilgartner, "Mapping Systems and Moral Order," 131.

31. National Cancer Institute, "An Assessment of the Impact of the NCI Cancer Biomedical Informatics Grid (caBIG®)."

32. For a more detailed analysis of both of these cases, see Leonelli, "Global Data for Local Science" and "What Difference Does Quantity Make?"

33. For a fascinating case of a data infrastructure whose origins date back to the 1950s and whose success has been marked by moments of failure and profound transformation, see the Long Term Ecological Research community (Aronova, Baker, and Oreskes, "Big Science and Big Data in Biology"; Baker and Millerand, "Infrastructuring Ecology").

34. Organisation for Economic Co-Operation and Development, "Guidelines for Human Biobanks and Genetic Research Databases (HBGRDs)"; Royal Society, "Science as an Open Enterprise."

35. Stemerding and Hilgartner, "Means of Coordination in Making Biological Science," 60.

36. These shifts in science policy are evidenced by the recommendations of the National Working Group on Expanding Access to Published Research Findings in the United Kingdom, also known as the Finch Report (Finch Group, "Accessibility, Sustainability, Excellence"), and the subsequent statement by Research Councils UK, "RCUK Policy on Open Access."

37. See for instance Martin, "Genetic Governance"; Gibbons, "From Principles to Practice"; Leonelli, Spichtiger, and Prainsack, "Sticks AND Carrots."

38. Christine Borgman and collaborators have long documented this shift of attitudes in the United States (e.g., Borgman, "The Conundrum of Sharing Research Data"); for an analysis of the evolving situation in the United Kingdom, where Open Data policies have been even more significantly supported, see Levin et al., "How Do Scientists Understand Openness?"

39. I participated in several such meetings and related "data science" training sessions in the United Kingdom, the Netherlands, and India, which made me aware of the extent of disagreement and confusion surrounding these key issues.

40. Mazzotti, "Lessons from the L'Aquila Earthquake."

41. Grundmann, "'Climategate' and the Scientific Ethos."

42. Royal Society, "Science as an Open Enterprise."

43. Rappert and Balmer, *Absence in Science, Security and Policy*; Leonelli, Rappert, and Davies, "Data Shadows: Knowledge, Openness and Access."

44. Bezuidenhout et al., "Open Data and the Digital Divide in Life Science Research."

45. For instance, GlaxoSmithKline is the main sponsor of the Target Validation Centre for biomedical data dissemination and analysis based at the European Bioinformatics Institute in Hinxton, Cambridge.

46. Fortun, *Promising Genomics*.

47. Tutton and Prainsack, "Enterprising or Altruistic Selves?"

48. Sunder Rajan, *Biocapital*; Kelty, *Two Bits*.

49. Bastow and Leonelli, "Sustainable Digital Infrastructure."

50. Elixir, "Elixir."

51. International Arabidopsis Informatics Consortium, "An International Bioinformatics Infrastructure to Underpin the *Arabidopsis* Community."

52. This conception of value builds specifically on the work of James Griesemer, who presented a paper on modes of attention at the first Knowledge/Value workshop in 2011 (documented on Knowledge/Value, "Concept Note for the Workshop Series"), and Sunder Rajan's contemporary reading of Marxist views on value as an "abstraction device" (*Pharmocracy*, chapter 1). More broadly, this way of thinking about value encompasses insights from philosophical literature concerned with values in science, as well as Science and Technology Studies scholarship looking at ways in which value is created, measured, and mobilized within and beyond knowledge-producing activities. In reference to the former approach, see Longino, *Science as Social Knowledge*; Wylie, Kinkaid, and Dupré, *Value-Free Science?*; and Douglas, *Science, Policy and the Value-Free Ideal*. For an example of the latter approach, see Dussauge, Helgesson, and Lee, *Value Practices in the Life Sciences and Medicine*.

53. Indeed, considering the affective value of data could be an effective way to confront the ethical issues surrounding big data and more generally the ethical and epistemological difficulties generated by the increasing disassociation between data and the people, materials, or processes from which such data are originally extracted.

54. Hinterberger and Porter, "Genomic and Viral Sovereignty."

55. Leonelli et al., "Making Open Data Work for Plant Scientists"; Levin, "What's Being Translated in Translational Research?"

56. I am grateful to Kaushik Sunder Rajan for helpful discussions on this key point. Some of our joint reflections on the intersections between knowledge and value have been published as Sunder Rajan and Leonelli, "Introduction."

57. GlaxoSmithKline, "Data Transparency."

58. Nisen and Rockhold, "Access to Patient-Level Data from GlaxoSmithKline Clinical Trials."

59. Of course, my brief analysis of scientific, political, financial, and affective value of data is not meant to be exhaustive, and much more could be said about cultural, social, and many other forms of value.

CHAPTER THREE

1. This approach is inspired by Theodore Porter's view of quantification as a "technology of distance" whose main function is to facilitate communication among sometimes far-flung interlocutors (Porter, *Trust in Numbers*).

2. This is what Hans-Jörg Rheinberger calls the "medial world of knowledge-making" ("Infra-Experimentality," 340).

3. Giere, *Scientific Perspectivism*. See also Gooding, *Experiment and the Making of Meaning*; and Radder, *The Philosophy of Scientific Experimentation* and "The Philosophy of Scientific Experimentation."

4. In the 1980s, several philosophers and philosophically minded historians and sociologists pioneered the study of the relation between observation and experiments (see, for instance, Latour and Woolgar, *Laboratory Life*; Hacking, *Representing and Intervening*; Collins, *Changing Order*; Franklin, *The Neglect of Experiment*; Galison, *How Experiments End*; Bogen and Woodward, "Saving the Phenomena"). In bringing the results of observation and experiments under the same umbrella, I follow the perspective offered by Radder, *The World Observed / The World Conceived*, according to which observations are as situated and subject-dependent as experimental results.

5. Hanson, *Patterns of Discovery*, 19.

6. See Bogen, "Noise in the World," 10; and Schindler, "Theory-Laden Experimentation," for discussions of different types of theory-ladenness. Of course, these views have a much longer and illustrious lineage than what I have briefly sketched here, including most prominently the work of Francis Bacon and William Whewell.

7. Science and Technology Studies scholars have long discussed the very idea of "raw data" as "bad" or even as an "oxymoron" (recent examples include Bowker, *Memory Practices in the Sciences*; Gitelman, *"Raw Data" Is an Oxymoron*). While I appreciate the careful studies of data production and maintenance within such scholarship, and their critical tone vis-à-vis hyped discourse on big data, I want to emphasize that at least some scientists (certainly in biology) consider the idea of "raw" data to be valuable and that this raises interesting questions around disciplinary and other sources of variability in how data are understood and handled. This point is also stressed by Paul Edwards, *A Vast Machine*, as I will discuss later in the chapter.

8. Hempel, "Fundamentals of Concept Formation in Empirical Science," 674.

9. In the words of James Woodward, this involved conceptualizing the evidential relation as "a purely formal, logical, or a priori matter" (Woodward, "Data, Phenomena, and Reliability," S172).

10. Reichenbach, *Experience and Prediction*.

11. Francis Bacon, *The Novum Organum*; William Whewell, "Novum Organon Renovatum, Book II"; and John Stuart Mill, *A System of Logic*.

12. Notable contributions include Cartwright, *How the Laws of Physics Lie*; Bechtel and Richardson, *Discovering Complexity*; Morgan and Morrison, *Models as Mediators*; Giere, *Scientific Perspectivism*; Weisberg, *Simulation and Similarity*; Bailer-Jones, *Scientific Models in Philosophy of Science*; de Chadaverian and Hopwood, *Models*; Creager, Lunbeck, and Wise, *Science without Laws*; and Morgan, *The World in the Model*, among many others.

13. See, in particular, Kellert, Longino, and Waters, *Scientific Pluralism*; and Chang, *Is Water H$_2$O?* As a result, some philosophers have become convinced that understanding scientific reasoning means studying the history and characteristics of actual research practices across different periods, locations, and disciplines (Ankeny et al., "Introduction: Philosophy of Science in Practice"; Chang, *Inventing Temperature*).

14. Hacking, "The Self-Vindication of the Laboratory Sciences," 48. Peter Galison has taken a similar position with respect to data obtained through experiments in particle physics, which has enabled him to study the ways in which data are exchanged across research communities in this area and how their scientific use is affected by their travels (*How Experiments End* and *Image and Logic*).

15. Rheinberger, "Infra-Experimentality," 6–7. Data may thus be viewed as resulting from processes of *material abstraction*, such as the ones I discussed elsewhere in relation to the production and use of model organisms (Leonelli, "Performing Abstraction").

16. Latour, "Circulating Reference."

17. See the overview of sequence data formats provided by the Computational Biology Research Group of the University of Oxford, "Examples of Common Sequence File Formats."

18. See for instance Fry, *Visualizing Data*. I should stress that Rheinberger, in *Towards a History of Epistemic Things* and "Infra-Experimentality," grounds his views on the study of sequencing techniques developed in the 1970s, which explains why he did not take subsequent developments in data production and analysis into account.

19. On iterativity, see Chang, *Inventing Temperature*; O'Malley, "Exploration, Iterativity and Kludging in Synthetic Biology"; and Wylie, *Thinking from Things*.

20. This is because, particularly as a result of crowdsourcing initiatives and "citizen science," individuals who do not have formal scientific training and are not professionally employed to conduct research can and do play a significant role in gathering, handling, and interpreting data for the purpose of knowledge production. I return to this point in chapter 6, section 1.3, and chapter 7.

21. This example takes inspiration from Richard Lenski's research program in experimental evolution (Blount et al., "Genomic Analysis of a Key Innovation in an Experimental *Escherichia coli* Population").

22. The definition of data provided by the Royal Society, which portrays them as "numbers, characters or images that designate an attribute of a phenomenon" ("Science as an Open Enterprise," 12), can also be interpreted in this vein.

23. This is not to say that a photograph of an embryo or a microarray slide could potentially be taken to function as evidence for any biological phenomenon—there are clearly limits to the number of interpretations that any specific dataset can be given, and these two forms of data will unavoidably capture some information about one embryo and a specific pattern of gene expression, respectively. At the same time, even the most specialized type of data can be used as evidence for claims about a variety of phenomena, depending on what other data and background assumptions it is integrated with and who makes the assessment. Thus, while a microarray slide will unavoidably provide information about gene expression, such information may be interpreted as evidence for a claim about the development of different phenotypic traits or different markers in the genome.

24. Another way to frame this point is to say that data have no fixed information content, since that content is itself a function of the material features of data and of the expertise and skills of whoever interprets them. I will not pursue this in detail because a thorough discussion of the relationship between data and information exceeds the scope of this book; here I focus on the artifacts that are taken to encode information—data—and the processes through which such attribution is made, rather than on the notion of information per se. It would be productive, however, to bring insights from the philosophy of science to bear on the burgeoning field of philosophy of information, which is displaying increasing attention to the semantic aspects of communication (see for instance Floridi and Illari, *The Philosophy of Information Quality*). A prominent contributor within that literature is Luciano Floridi, whose account of the status and role of data bears strong similarities to the one I presented here. We both stress that data are a relational category, which can be clearly distinguished from the categories of metadata and model, and that a credible philosophical account needs to take account of the wide variety of situations in which data may be located, as well as the impact that such location has on what counts as data in the first place (*The Philosophy of Information*, 85). Floridi also proposes that "there can be no information without data representation" ("Is Information Meaningful Data?"), an insight that mirrors my own position on the relation between data and models (see section 3.4). One point of disagreement concerns the truth-value of data, which Floridi thinks can be assessed independently of how they are used: they can be inaccurate, imprecise or incomplete—untruthful—regardless of their location and handling (*The Philosophy of Information*, 104). By contrast, I maintain that data do not have truth-value in and of themselves but only in relation to specific claims and theoretical, material, and social conditions.

25. Shapin Schaffer, *Leviathan and the Air-Pump*.

26. For instance, three biologists gathering around a Petri dish to witness the movement of a slime mold, as in John Bonner's seminal experiments; see Princeton University, "John Bonner's Slime Mold Movies."

27. This insight links my approach to that proposed by Lorraine Daston and Peter Galison, who described data as portable, workable, and "communal," and observed how these characteristics define what data are and how they function

across research contexts (Daston and Galison, "The Image of Objectivity"), and by Mary Morgan, who also stressed the crucial importance of movement to assessing the value of data as evidence ("Travelling Facts," *The World in the Model*). The cases of data travel examined by these authors, in conjunction with my own study of the complexity of the decontextualization processes involved in data curation, clearly demonstrate the diverse ways in which data can be made to travel and the importance of acknowledging efforts to make data portable as a scientific achievement in its own right.

28. Of course, data in a digital format are only visible and manipulable via interaction with computer screens, and the physical constraints and the type of resistance offered by virtual environments are different from those encountered in nonvirtual situations. For my purposes, whether data come in digital or analog form matters insofar as it determines the ways in which data can be disseminated. Yet again, the extent to which this affects data reuse depends on the specific case, especially given the ease with which data can change from analog to digital and vice versa (e.g., printing out a paper copy of data viewed on a computer screen is a relatively easy task in the case of a small data sample but becomes impossible when involving large datasets such as whole genome sequences; similarly, scanning and annotating a photograph for insertion into a digital database may be straightforward or extremely complex depending on how the database is set up). This is a different stance from the one taken by Morgan in her explorations of "experiments without material intervention" ("Experiments without Material Intervention"). What I am interested in is not the difference between the kinds of interventions involved in in vivo versus in silico research but rather the ways in which information acquired through in vivo experiments is treated (including its modeling and dissemination in silico). In this sense, my position is closer to Wendy Parker's idea of "computer experiments as, first and foremost, experiments on real material systems" ("Does Matter Really Matter?" 488).

29. Orlin Vakarelov provides a useful characterization of a medium as "the concrete stuff. It is the system that gets pushed and pulled by the rest of the world . . . in just the right way to support the patterns of interactions that constitute the information process" (Vakarelov, "The Information Medium," 49). Vakarelov's general theory of information media is complementary to the account I am providing here, particularly when he stresses that information needs to be understood "as a phenomenon in the interaction and coordination of large networks of information media, implemented in the highly organized dynamics of the world, and interacting with the world at a causal level" (ibid., 65).

30. See, for instance, Michael Ghiselin, "A Radical Solution to the Species Problem," and David Hull "Are Species Really Individuals?" A more recent example is the work by John Dupré and Maureen O'Malley on the philosophy of microbiology and processual approaches to life (Dupré and O'Malley, "Metagenomics and Biological Ontology"; Dupré, *Processes of Life*; O'Malley, *Philosophical Issues in Microbiology*).

31. Guay and Pradeau, *Individuals Across the Sciences*.

32. This view parallels James Griesemer's discussion of data production and collection as tracking activities, whose material result is the generation of objects that can be treated as documents of the phenomena being tracked. This is compat-

ible with a relational understanding of what counts as data, particularly given its
emphasis on the role of scientists' commitments—whether practical (to instruments and materials), social (to networks, peers, and institutions) or theoretical (to
specific conceptualizations and background knowledge)—in conducting tracking activities and interpreting their products (Griesemer, "Reproduction and the
Reduction of Genetics," "Theoretical Integration, Cooperation, and Theories as
Tracking Devices," and "Formalization and the Meaning of 'Theory' in the Inexact
Biological Sciences").

33. This objection was first raised to me by Chris Timpson, who discusses it
in Chapter 2 of *Quantum Information Theory and the Foundations of Quantum
Mechanics*.

34. Biagioli, "Plagiarism, Kinship and Slavery."

35. Does that mean that every time a dataset is copied or translated into a new
format, its scientific significance necessarily changes? I do not think so, because
in my framework it is not a change in format per se that determines changes in
the scientific significance of data but rather a change of situation of inquiry. It is
therefore perfectly reasonable for a community of biologists to treasure the idea
that "raw data" can be disseminated with minimal modifications across labs. The
ways in which a dataset is interpreted depend both on its location and on its format, and which of these factors affects data reuse is a matter of local assessment.
This observation cuts through the philosophical distinctions between token and
type, immaterial form (idea) and tangible work (objects). The need to recognize the
epistemological significance of the variability and polymorphism involved in data
journeys is why I insist on avoiding a definition of data that would make it possible
to identify and discuss them independently of a specific context. There is no such
thing as data in and of themselves, as what counts as data is always relative to a
given inquiry where evidence is sought to answer, or even formulate, a question.
Data are not only modifiable in principle but are in fact frequently modified during
their travels in ways that profoundly affect their ability to function as evidence.

36. Woodward, "Phenomena, Signal, and Noise."

37. Bogen and Woodward, "Saving the Phenomena," 314.

38. Indeed, data "cannot be produced except through processes whose results
reflect the influence of causal factors that are too numerous, too different in kind,
and too irregular in behavior for any single theory to account for them" (Bogen,
"Noise in the World," 18).

39. Bogen and Woodward, "Saving the Phenomena," 306.

40. Significant discussions of the data-phenomena distinction, which also ques-
tioned the status of data, include McAllister, "Phenomena and Patterns in Data
Sets," "Model Selection and the Multiplicity of Patterns in Empirical Data," "The
Ontology of Patterns in Empirical Data," and "What Do Patterns in Empirical
Data Tell Us About the Structure of the World?"; Schindler, "Bogen and Wood-
ward's Data/Phenomena Distinction"; Massimi, "From Data to Phenomena"; and
Teller, "Saving the Phenomena Today."

41. Woodward, "Phenomena, Signal, and Noise," 792. Notably, in the same
paragraph, Woodward goes on to say that "by extension, records or reports of such
outcomes may also be regarded as data." This seems to indicate that Woodward
recognizes the difficulties involved in restricting the notion of data to "outcomes of

measurement or detection processes"; however, neither Woodward nor Bogen have devoted more attention to this issue.

42. Bogen and Woodward, "Saving the Phenomena," 317.

43. It should thus be clear that my account does not consider nonlocality as involving a complete decoupling between data and their production context; rather, I take this term to indicate the ability of data to be adopted in contexts other than the one in which they have been originated—which, as I show, requires a critical assessment of metadata documenting their provenance.

44. Note that this view does not necessarily involve a relativist and/or constructivist perspective on phenomena, such as that presented by James McAllister ("Model Selection and the Multiplicity of Patterns in Empirical Data," "The Ontology of Patterns in Empirical Data," and "What Do Patterns in Empirical Data Tell Us About the Structure of the World?"). Rather, my conceptualization of phenomena follows Michela Massimi's Kantian stance, according to which "phenomena are (partially) constituted by us, rather than being ready-made in nature" (Massimi, "From Data to Phenomena," 102). "Partially" is a crucial qualification in this view. While I am sympathetic to McAllister's portrayal of phenomena as "labels that investigators apply," I do not support his conclusion that these labels can be applied "to whichever patterns in data sets [investigators] wish so to designate" (McAllister, "Phenomena and Patterns in Data Sets," 224). This kind of constructivism dismisses the significance of the material conditions under which judgments about labels and classifications are made. As I illustrate in chapter 5, the classification of data is not arbitrary, but rather it is strongly constrained by the conditions under which humans interact with the world. The identification of phenomena is thus both the result of conceptual judgments by investigators and the outcome of complex material and social assemblages of data production and dissemination, which are not determined by any one specific conceptualization of which phenomena exist and can be investigated.

45. Bogen and Woodward, "Saving the Phenomena," 326.

46. I discuss these characteristics, and the relation between claims and data, in more detail in chapter 5.

47. I discussed big data in biology in Leonelli, "What Difference Does Quantity Make?"

48. Fry, *Visualizing Data*; Hey, Tansley, and Tolle, *The Fourth Paradigm*.

49. Suppes, "Models of Data," 258. Suppes first articulated this idea in relation to data underlying learning theory in psychology and later brought his ideas to bear also on cases from physics and astronomy ("From Theory to Experiment and Back Again," "Statistical Concepts in Philosophy of Science").

50. Suppes presents a hierarchy of theories, models, and problems plaguing scientific research, at the bottom of which he places "every intuitive consideration of experimental design that involves no formal statistics" ("Models of Data," 258; see Suppes, "Statistical Concepts in Philosophy of Science," for a general analysis of pragmatic aspects of experimentation).

51. Suppes was actually interested in large-scale computation and the widespread availability of information and argued that the shift in speed of data retrieval would "change the world" ("Perception, Models and Data," 112; see also Suppes, "The Future Role of Computation in Science and Society"). However, he

did not explicitly analyze how this argument would reflect on conceptualizations of data themselves; and his careful studies of data processing did not extend outside of the laboratory to include dissemination tools such as databases.

52. Frigg and Hartmann, "Models in Science."

53. Edwards, *A Vast Machine*, 280.

54. Ibid, 272.

55. Edwards, *A Vast Machine*, 291.

56. This characterization is broadly aligned with existing accounts of the role of models in biological practice, such as Michael Weisberg's view of models as "potential representations of target systems" (*Simulation and Similarity*, 171) and Miles MacLeod and Nancy Nersessian's description of model-based reasoning as "a process of generating novel representations through the abstraction and integration of constraints from many different contexts" ("Building Simulations from the Ground Up," 534).

57. The notion of representation employed here does not require isomorphism, and even straightforward similarity, between a model and a target system—in this sense, I follow Morgan's account of modeling as "the activity of creating small worlds expressed in another medium" (Morgan, *The World in the Model*, 30), where the extent to which such small worlds relate to the ideas of their creators (theorize) vis-à-vis the actual structure of the world (describe) varies in each specific case.

58. Morgan highlighted this point in her review of philosophical literature on model development, where she notes that no matter which account we follow, modeling always involves "the scientists' intuitive, imaginative, and creative qualities" (ibid., 25). Strikingly, Morgan does not mention the word "data" once in her account. This makes perfect sense when one considers the concrete differences, rather than the possible overlaps, between the activities of model making and data packaging.

59. I discussed this performative understanding of abstraction in Leonelli, "Performing Abstraction." Ulrich Krohs and Werner Callebaut provided a vivid illustration of this process in their discussion of the relation between models and data in systems biology, where they point out that the first step toward building models of data is typically to "single out a metabolic or sensory capacity"—in other words, a feature of the biological world—"that is to be analyzed and eventually modelled" ("Data without Models Merging with Models without Data," 188).

60. Borgman, "The Conundrum of Sharing Research Data," 1061.

CHAPTER FOUR

1. I discussed this notion in more detail in Leonelli, *Weed for Thought*, and "The Impure Nature of Biological Knowledge."

2. Ryle's view on embodied knowledge stems from his distinction between "knowing how" and "knowing that" (*The Concept of Mind*). My characterization of embodied knowledge is also closely aligned with Hans Radder's detailed account of what the performance of an experiment entails in *The Material Realization of Science* as well as Ken Waters's discussion in "How Practical Know-How

Contextualizes Theoretical Knowledge" of what he calls procedural knowledge, particularly his analysis of the intertwining of explanatory reasoning and investigative practices in classical genetics.

3. Polanyi, *Personal Knowledge*, 54–55.

4. Ibid., 55.

5. Ryle, *The Concept of Mind*, 30.

6. Ibid., 42.

7. Ibid., 29.

8. Many scholars in science and technology studies have documented these processes, for example, Latour and Woolgar, *Laboratory Life*; and Michael Lynch, "Protocols, Practices, and the Reproduction of Technique in Molecular Biology."

9. JoVE, "About."

10. Taylor et al., "Promoting Coherent Minimum Reporting Guidelines for Biological and Biomedical Investigations."

11. Avraham et al., "The Plant Ontology Database."

12. BioSharing, "Homepage."

13. Keller, *A Feeling for the Organism*.

14. Edwards et al., "Science Friction."

15. Personal correspondence, November 2009.

16. Rosenthal and Ashburner, "Taking Stock of Our Models"; Leonelli, "Growing Weed, Producing Knowledge."

17. The reverse can also be the case: for instance, I have documented the case of C24, which was erroneously treated by experimenters as an ecotype of *Arabidopsis* until the mistake was spotted by stock center technicians (Leonelli, "Growing Weed, Producing Knowledge," 214–15).

18. See, for example, Rader, *Making Mice*; Davies, "What Is a Humanized Mouse?" and "Arguably Big Biology."

19. Love and Travisano, "Microbes Modeling Ontogeny"; O'Malley, *Philosophical Issues in Microbiology*.

20. See CASIMIR website, http://casimir.org.uk.

21. Draghici et al., "Reliability and Reproducibility Issues in DNA Microarray Measurements."

22. Leonelli, "When Humans are the Exception."

23. Brazma et al., "Minimal Information About a Microarray Experiment (MIAME): Towards a Standard for Microarray Data."

24. Shields, "MIAME, We Have a Problem"; Rogers and Cambrosio, "Making a New Technology Work."

25. McCarthy et al., "Genome-Wide Association Studies for Complex Traits."

26. See, for example, McMullen, Morimoto, and Nunes Amaral, "Physically Grounded Approach for Estimating Gene Expression from Microarray Data."

27. Leonelli et al., "Making Open Data Work for Plant Scientists."

28. See, for example, Fernie et al., "Recommendations for Reporting Metabolite Data."

29. I am here deliberately avoiding the strict distinction between experimenters and technicians employed by Hans Radder in his account of material realization. While Radder uses it effectively to remind readers of the importance of division of labor within laboratories, he also posits that technicians act as the main reposito-

ries of embodied knowledge and characterizes experimenters as mostly interested in obtaining theoretical (propositional) knowledge. This is not what I have observed in biological laboratories, where technicians are certainly more au fait with the functioning of specific machines (e.g., a spectrometer), but experimenters are also deeply invested in the material realization of experiments, including the development of new tools and techniques and the handling of organic specimens. The importance of embodied knowledge in experimental work has been stressed by several commentators, including Gooding, *Experiment and the Making of Meaning*; Franklin, *The Neglect of Experiment*; and Rheinberger, "Infra-Experimentality." The celebrated, twice-Nobel-laureate Fred Sanger put it this way: "Of the three main activities involved in scientific research, thinking, talking and doing, I much prefer the last and am probably best at it" (Sanger, "Sequences, Sequences and Sequences," 1).

30. Christine Hine captured the broader sense in which several types of experts, networks, and regulatory structures actively participate in the development of biological databases with the expression "dance of initiatives" (Hine, "Databases as Scientific Instruments and Their Role in the Ordering of Scientific Work").

31. This is highlighted by both JoVE and BioSharing on the homepage of their websites, on which they proudly declare to be "part of the growing movement for reproducible research" (accessed March 19, 2014).

32. Parker, Vermeulen, and Penders, *Collaboration in the New Life Sciences*; Davies, Frow, and Leonelli, "Bigger, Faster, Better?"

33. Goble and Wroe, "The Montagues and the Capulets"; Searls, "The Roots of Bioinformatics"; Stevens, *Life Out of Sequence*.

34. Callebaut, "Scientific Perspectivism"; Chang, *Is Water H_2O?*; Giere, *Scientific Perspectivism*; Longino, *The Fate of Knowledge*; Mitchell, *Biological Complexity and Integrative Pluralism*.

35. Chang, *Is Water H_2O?*, 272.

36. Ibid., 151.

37. Giere, *Scientific Perspectivism*.

38. This view has strong connections to the idea of distributed cognition famously championed by Edward Hutchins in his study of collective agency in ship navigation (*Cognition in the Wild*), an idea that Andy Clark extends to all cases where "computational power and expertise is spread across a heterogeneous assembly of brains, bodies, artifacts, and other external structures" (*Being There*, 77). I plan to expand on these parallels in future work.

39. Such as a public indictment for fraud, as in the case of Korean stem cell researcher Hwang Woo-suk (Hong, "The Hwang Scandal that 'Shook the World of Science'"), or failure of reproduction attempts, such as that revealed in psychology experiments (Open Science Collaborative, "Estimating the Reproducibility of Psychological Science").

40. For instance, data inferred by direct assay is typically ranked higher than data inferred by electronic annotation or computational prediction.

41. For instance, assessing every step of the journey undertaken by one microarray dataset within The Arabidopsis Information Resource involves expertise in microarray experiments on plants, as well as the use of specific software, the choice of bio-ontologies and metadata, and the ways in which microarray visu-

alizations can be instantiated within the database infrastructure, all of which are unlikely to be possessed by a single individual—and this is a relatively simple case of data travel!

42. For an examination of the notion of understanding and its relation to embodied knowledge, see Leonelli, "The Impure Nature of Biological Knowledge;" for a forceful defense of the interrelations between social and cognitive dimensions of scientific knowledge, see Longino, *The Fate of Knowledge*.

43. Helmreich, *Silicon Second Nature*; King et al., "The Automation of Science."

44. Evans and Rzhetsky, "Machine Science."

45. See, for example, Anderson, "The End of Theory."

46. I am here following Radder's formulation of the distinction between experimental reproducibility and replicability (in *The Material Realization of Science*), where the former denotes the ability to obtain the same outputs by a variety of means, while the latter refers to the ability to carry out the same experimental procedures to the same effect.

47. Casadevall and Fang, "Reproducible Science"; Mobley et al., "A Survey on Data Reproducibility in Cancer Research Provides Insights into Our Limited Ability to Translate Findings from the Laboratory to the Clinic." For a conceptual analysis of this problem, see Harry Collins's seminal discussion of the rarity of experimental replication and the problem of experimenters' regress (*Changing Order*).

CHAPTER FIVE

1. O'Malley and Soyer, "The Roles of Integration in Molecular Systems Biology."

2. Rasmus Winther, "The Structure of Scientific Theories," provides a comprehensive discussion of the pragmatic view of theories.

3. Giere, *Science without Laws* and *Scientific Perspectivism*.

4. Suárez and Cartwright, "Theories: Tools versus Models."

5. Knorr Cetina, *Epistemic Cultures*.

6. Stotz, Griffiths, and Knight, "How Biologists Conceptualize Genes."

7. See, for example, Latour, "Circulating Reference"; Bowker and Star, *Sorting Things Out*.

8. Brazma, Krestyaninova, and Sarkans, "Standards for Systems Biology," 595.

9. Lewis, "Gene Ontology," 103.1.

10. In chapter 2, I showed how the success of this approach was due as much to its technical characteristics as to its institutional implementation. The development of this system required the creation of a new set of organizations bringing together researchers from different communities and serving as an interface between bioinformaticians, researchers using data, and regulatory bodies such as funding agencies and international organizations.

11. In "Documenting the Emergence of Bio-Ontologies," I have shown how the term "ontology" is used in computer science to indicate a set of representational primitives with which to model a domain of knowledge or discourse. It was originally picked up from philosophy, but it quickly acquired a technical meaning

that had everything to do with the structure of "object-directed" programming languages and little to do with the intricacies of metaphysics (Gruber, "A Translation Approach to Portable Ontology Specifications"). Some philosophers, most notably Barry Smith, have exploited the overlap in terminology so as to position metaphysics as an important source of inspiration and guidance in the construction of bio-ontologies (e.g., Smith and Ceusters, "Ontological Realism"). Smith effectively brought his own philosophical viewpoint to bear on the nonphilosophical meaning assigned to the term by bioinformaticians, and his work is partly responsible for the interdisciplinary mingling that underpins the "mixed status" of the term "ontology." I will not examine Smith's views in detail in this book. For my current purposes, it suffices to note that contributions like Smith's have successfully and fruitfully shaped the development of bio-ontologies without, however, engaging in an epistemological analysis of the implications of applying metaphysical principles for the future of biological research. My analysis of bio-ontologies as theories hopefully contributes to understanding how the adoption of general metaphysical principles to organize data journeys can affect the development of scientific knowledge.

12. Brazma, Krestyaninova, and Sarkans, "Standards for Systems Biology," 601.

13. Bard and Rhee, "Ontologies in Biology," 213.

14. Ashburner et al., "Gene Ontology," 27.

15. All three GO ontologies are designed to represent the processes, functions, and components of a generic eukaryotic cell; at the same time, they can incorporate organism-specific features (in 2012, GO already included data from over thirty species). See Ashburner et al., "Gene Ontology," and the Gene Ontology "Minutes" from 2004, 2006, and 2007.

16. Note that "regulates" is not organized in the parent/child structure. Also, in 2013, the Gene Ontology was in the process of incorporating another two types of relations, "has_part" and "occurs_in," and could potentially adopt many other types of relations, as documented by Smith et al., "Relations in Biomedical Ontologies."

17. See for example Liang et al., "Gramene."

18. Smith et al., "The OBO Foundry."

19. It is important to note the specific focus of my discussion at the outset, because since the development of GO, several other formalizations for bio-ontologies were introduced and the status and place of the OBO format among them is hotly disputed (Egaña Aranguren et al., "Understanding and Using the Meaning of Statements in a Bio-Ontology"). For instance, ontologies constructed using the Web Ontology Language (OWL) are attracting attention as a useful alternative to the OBO format, and efforts are under way to make the two systems compatible (e.g., Hoehndorf et al., "Relations as Patterns"; Tirmizi et al., "Mapping between the OBO and OWL Ontology Languages"). These differences and convergences are crucial to the future development of bio-ontologies and the extent of their popularity in model organism research; however, they are not relevant to my philosophical analysis, which focuses on the features currently adopted by the OBO consortium to serve the community of experimental biologists that uses them.

20. Keet, "Dependencies between Ontology Design Parameters."

21. Ashburner et al., "Gene Ontology"; Lewis, "Gene Ontology"; Renear and Palmer, "Strategic Reading."

22. Bada et al., "A Short Study on the Success of the Gene Ontology."

23. Baclawski and Niu, *Ontologies for Bioinformatics*, 35.

24. Gene Ontology Consortium, "GO:0046031."

25. This of course does not mean that they always are updated in this way. As I discussed in chapter 2, database maintenance in the long term is a labor-intensive process that cannot happen without adequate sponsorship. Nevertheless, recent research shows remarkably stable levels of updating activity within major bio-ontologies (Malone and Stevens, "Measuring the Level of Activity in Community Built Bio-Ontologies").

26. Bada et al., "A Short Study on the Success of the Gene Ontology," 237. I reviewed instances of GO adapting to changes in biological knowledge in Leonelli et al., "How the Gene Ontology Evolves."

27. Kuhn, *The Structure of Scientific Revolutions.*

28. Hesse, *Revolutions and Reconstructions in the Philosophy of Science*, 108.

29. Ibid., 97.

30. Personal correspondence, August 27, 2004.

31. Hempel, *The Structure of Scientific Inference*, 4–26; and Hesse, *Revolutions and Reconstructions in the Philosophy of Science*, 84.

32. Gene Ontology Consortium. "Minutes of the Gene Ontology Content Meeting (2004)."

33. Paracer and Ahmadjian, *Symbiosis.*

34. Bernard, *An Introduction to the Study of Experimental Medicine*, 165.

35. In a similar vein, Kenneth Schaffner has offered a sophisticated view of "middle-range theories" as devices for knowledge representation and discovery, in which he even briefly explored the usefulness of object-oriented approaches to programming (the ancestors of bio-ontologies) as a way of expressing theoretical commitments (Schaffner, *Discovery and Explanation in Biology and Medicine*). I see his approach as congenial to mine, yet I am not analyzing it in detail here because of his different emphasis and targets (the relation of law-like statements in biology to those in physics and chemistry).

36. Dupré, *The Disorder of Things* and "In Defence of Classification"; and Dupré and O'Malley, "Metagenomics and Biological Ontology."

37. Dupré, "In Defence of Classification."

38. Dupré and O'Malley, ibid.

39. Müller-Wille, "Collection and Collation"; Müller-Wille and Charmantier, "Natural History and Information Overload."

40. Strasser, "The Experimenter's Museum."

41. Love, "Typology Reconfigured," 53.

42. Dupré, *The Disorder of Things* and "In Defence of Classification"; Brigandt, "Natural Kinds in Evolution and Systematics"; Reydon, "Natural Kind Theory as a Tool for Philosophers of Science."

43. Minelli, *The Development of Animal Form.*

44. Love, "Typology Reconfigured," 63; see also Griesemer, "Periodization and Models in Historical Biology."

45. Klein and Lefèvre, *Materials in Eighteenth-Century Science.*

46. Daston, "Type Specimens and Scientific Memory," 158.

47. Crombie, *Styles of Scientific Thinking in the European Tradition*; and Hacking, *Historical Ontology.*

48. Pickstone, *Ways of Knowing*, 60.

49. For a congenial account of the role of theories as mechanisms for domain control, whose formalization is a way to delimit what counts as relevant objects, techniques, and instruments in biology, see Griesemer, "Formalization and the Meaning of 'Theory' in the Inexact Biological Sciences."

50. Leonelli et al., "How the Gene Ontology Evolves."

51. Richardson et al., "There Is No Highly Conserved Embryonic Stage in the Vertebrates."

52. Wimsatt and Griesemer, "Reproducing Entrenchments to Scaffold Culture." See also Caporael, Griesemer, and Wimsatt, *Developing Scaffolds in Evolution, Culture, and Cognition.*

53. The importance of models to understanding theories has been widely discussed, for instance, by Griesemer (e.g., "Theoretical Integration, Cooperation, and Theories as Tracking Devices") and by contributors to de Regt, Leonelli, and Eigner, *Scientific Understanding.*

54. Krakauer et al., "The Challenges and Scope of Theoretical Biology," 272.

55. The labels used in classificatory theories may be viewed as *concepts* in the sense outlined by Hans Radder (*The World Observed / The World Conceived*): they have nonlocal meaning that can be applied in a variety of situations and yet are the result of a process of abstraction from specific research circumstances.

56. This situation has been highlighted both by Nancy Cartwright's work on low-level regularities applying to observational entities (*How the Laws of Physics Lie* and *The Dappled World*) and Griesemer's analysis of formalization ("Formalization and the Meaning of 'Theory' in the Inexact Biological Sciences").

57. One could question the extent to which this generalization has been successfully achieved. While I do not see this as crucial to my argument, which concerns the underlying aspiration of this system to generalize rather than its success in doing so, I discussed the successes and difficulties of this enterprise in Leonelli et al., "How the Gene Ontology Evolves"; and Leonelli, "When Humans Are the Exception."

58. Mitchell, *Unsimple Truths*, 50–51.

59. In Mitchell's words again: "Our philosophical conception of the character of scientific knowledge should be based on what types of claims function reliably in scientific explanations and predictions. Some generalizations are more like the strict model of laws; others are not. Both should be called laws. The more inclusive class of regularities, which I call pragmatic laws, are widespread in biology and, arguably, beyond" (Mitchell, *Unsimple Truths*, 51).

60. Cartwright, *How the Laws of Physics Lie.*

61. Scriven, "Explanation in the Biological Sciences."

62. Callebaut, "Scientific Perspectivism."

63. Anderson, "The End of Theory"; see also discussions by Callebaut, "Scientific Perspectivism"; and Wouters et al., *Virtual Knowledge.*

64. For example, Allen, "Bioinformatics and Discovery"; Kell and Oliver, "Here

Is the Evidence, Now What Is the Hypothesis?"; O'Malley et al., "Philosophies of Funding."

65. Waters, "The Nature and Context of Exploratory Experimentation."

66. Hill et al., "Gene Ontology Annotations."

67. This view parallels the endorsement of scientific perspectivism in the analysis of data-intensive science presented by Callebaut, "Scientific Perspectivism." Here I will not dwell on the thorny problem of scientific realism but rather point to the version in Chang's *Is Water H₂O?*, aptly called "active," which I think can account for the fallible and yet informative nature of propositional knowledge expressed in classificatory theories.

68. Star, "The Ethnography of Infrastructure."

CHAPTER SIX

1. O'Malley and Soyer, "The Roles of Integration in Molecular Systems Biology."

2. Historical work like Harwood, *Europe's Green Revolution and Others Since*; Müller-Wille, "Collection and Collation"; Smocovitis, "Keeping up with Dobzhansky"; and Kingsbury, *Hybrid*, clearly show the key role played by plant scientists in the development of several branches of biology, including evolutionary theory and genetics.

3. Leonelli, "Growing Weed, Producing Knowledge."

4. Browne, "History of Plant Sciences"; Botanical Society of America, "Botany for the Next Millennium."

5. Rhee, "Carpe Diem"; Koornneef and Meinke, "The Development of *Arabidopsis* as a Model Plant."

6. Of course, this does not mean that plant science is a homogeneous field or that there are no tensions and nonoverlapping programs at play within it. A major problem is the historical separation between agricultural and molecular approaches to plant science, which came into effect in the second half of the twentieth century (Leonelli et al., "Under One Leaf"). I also do not mean to assert that there are no comparable examples of cooperation in other fields; for analysis of such cooperation surrounding model organisms, see Kohler, *Lords of the Fly*; or Leonelli and Ankeny, "Re-Thinking organisms."

7. International Arabidopsis Informatics Consortium, "Taking the Next Step."

8. See, for example, Kourany, *Philosophy of Science after Feminism.*

9. This analysis is thus aligned with other attempts at broadening the notions of scientific research, collaboration, and knowledge traditionally supported within the philosophy of science, including, for instance, Longino, *The Fate of Knowledge*; Douglas, *Science, Policy, and the Value-Free Ideal*; Mitchell, *Unsimple Truths*; Elliott, *Is a Little Pollution Good for You?*; Nordmann, Radder, and Schiemann, *Science Transformed?*; and Chang, *Is Water H₂O?*.

10. Maher, "Evolution"; Behringer, Johnson, and Krumlauf, *Emerging Model Organisms* (vols. 1 and 2).

11. Spradling et al., "New Roles for Model Genetic Organisms in Understanding and Treating Human Disease."

12. Sommer, "The Future of Evo-Devo."

13. Bevan and Walsh, "Positioning Arabidopsis in Plant Biology."

14. I here adopt a broad definition of "level of organization," which is meant to reflect a way to organize and subdivide research topics that is still very popular within biology—that is, the focus on specific components and "scales" of biological organization, each of which requires a specific set of methods and tools of investigation that suit the dimensions and nature of the objects at hand.

15. Leonelli et al., "Under One Leaf."

16. See, for example, Stitt, Lunn, and Usadel, "*Arabidopsis* and Primary Photosynthetic Metabolism." See also Bechtel, "From Molecules to Behavior and the Clinic"; and Brigandt, "Systems Biology and the Integration of Mechanistic Explanation and Mathematical Explanation."

17. What I here call interlevel integration has been discussed at length by proponents of mechanistic explanation (Craver, "Beyond Reduction"; Darden, "Relations Among Fields"; Bechtel, "From Molecules to Behavior and the Clinic").

18. Bult, "From Information to Understanding"; Leonelli and Ankeny, "Re-Thinking organisms."

19. Leonelli et al., "Under One Leaf."

20. This view is compatible with an emphasis on experimental intervention ("learning by doing") and exploratory research as means to achieve biological discoveries (e.g., Steinle, "Entering New Fields"; Burian, "Exploratory Experimentation and the Role of Histochemical Techniques in the Work of Jean Brachet, 1938–1952"; O'Malley, "Exploratory Experimentation and Scientific Practice"; and Waters, "The Nature and Context of Exploratory Experimentation").

21. This scholarship includes Lindley Darden and Nancy Maull's classic paper "Interfield Theories" and, more recently, Bechtel and Richardson, *Discovering Complexity*; Mitchell, *Biological Complexity and Integrative Pluralism* and *Unsimple Truths*; and Brigandt, "Beyond Reduction and Pluralism" and "Integration in Biology."

22. As I discussed in chapter 4, comparisons between strains and varieties regarded as belonging to the same species can often be as interesting as comparisons among species. My analysis here is thus not meant to endorse strict taxonomic distinctions but rather to capture the importance of comparing differences between groups of organisms. Understood in this broad sense, "cross-species integration" can also include cross-variety and cross-strain analysis.

23. Babcock, *The Impact of US Biofuel Policies on Agricultural Price Levels and Volatility*.

24. Jensen, "Flowering Time Diversity in *Miscanthus*."

25. Jensen et al., "Characterization of Flowering Time Diversity in *Miscanthus* Species."

26. International Arabidopsis Informatics Consortium, "Taking the Next Step."

27. Gene Ontology Consortium, "AmiGO 2."

28. Leonelli et al., "How the Gene Ontology Evolves."

29. Quirin et al., "Evolutionary Meta-Analysis of Solanaceous Resistance Gene and *Solanum* Resistance Gene Analog Sequences and a Practical Framework for Cross-Species Comparisons."

30. Potter et al., "Learning From History, Predicting the Future."

31. Interestingly, the development of PlantPatho itself has involved considerable integrative efforts, both interlevel (by integrating data from different features of both host plants and pathogens) and cross-species (by integrating data coming from a variety of different species, though it must be noted that *Arabidopsis* research has provided much of the data used to start this initiative; Bülow, Schindler, and Hehl, "PathoPlant®").

32. It is important to stress here that human health can hardly be construed separately from the health of the environments that support human survival and well-being, particularly plants as key sources of air, fuel, and nurture. I thus endorse the idea of green biotechnology, and plant science, as playing a key role in protecting and improving human health. This may be perceived as counterintuitive by scholars who view red biotechnology, particularly the medical sciences, as the only form of knowledge that is concerned with human health; I hope that this analysis helps to correct this common misconception.

33. O'Malley and Stotz, "Intervention, Integration and Translation in Obesity Research."

34. This choice of priorities often comes at the expense of time dedicated to exploratory research and yet this does not necessarily compromise the quality of research and its outcomes. On the contrary, the development of research strategies to pursue socially relevant understandings of organisms, especially when it is coupled with the awareness that achieving such goals can sometimes be a long-term and complex endeavor, is a form of inquiry that philosophers of science should value and support. In this sense, my approach is closely aligned with the socially relevant philosophy of science proposed by Helen Longino (*The Fate of Knowledge*), Janet Kourany (*Philosophy of Science after Feminism*), and other leading philosophers interested in feminist approaches and social studies of science as key sources of insight for the development of the philosophy of science. See, for instance, the *Synthese* special issue edited by Fehr and Plaisance, "Making Philosophy of Science More Socially Relevant," and the review symposium on Kourany's *Philosophy of Science after Feminism* in *Perspectives on Science* ("Perspectives on Science After Feminism").

35. De Luca et al., "Mining the Biodiversity of Plants," 1660.

36. The "water-food-energy" nexus, as defined by the World Economic Forum, "Global Risks 2011: Sixth Edition," 8.

37. This is something that many funding bodies, particularly those subscribing to the "Impact Agenda" in the United Kingdom and elsewhere, are reluctant to acknowledge.

38. Bastow and Leonelli, "Sustainable Digital Infrastructure"; and Leonelli et al., "Making Open Data Work for Plant Scientists."

39. Pollock et al., *The Hidden Digital Divide*.

40. Bezuidenhout et al., "Open Data and the Digital Divide in Life Science Research."

41. Krohs, "Convenience Experimentation."

42. See Prainsack and Leonelli, "To What Are We Opening Science?"

43. See the historical studies by November, *Biomedical Computing*; García-Sancho, *Biology, Computing, and the History of Molecular Sequencing*; and Stevens, *Life Out of Sequence*.

44. Agar, "What Difference Did Computers Make?" 872.

45. Kohler, *Lords of the Fly*.

46. In *Life Out of Sequence*, Stevens interprets the unique fit between sequence analysis and existing computer facilities as a sign that computers have not adapted to biological research, but rather biologists have increasingly tailored their work to the opportunities offered by computing technologies. While agreeing with him on the importance of molecular research in brokering the marriage of biology and computing, I think that biological insights and demands have played—and continue to play—a more active role in shaping computing technology. Examples of this are the connectivist paradigm in robotics, where shifting understandings of cognition have occasioned new ways of conceptualizing hardware; research on organismal evolution and development, which has provided precious insights into how software and algorithms could develop; forms of synthetic biology where molecular mechanisms are being mimicked in virtual environments and organic tissues are themselves used as material platforms ("wetware") for computing; and, more generally, the biological need to integrate multiple data types for a variety of uses, which has pushed the introduction of innovations such as bio-ontologies and complex databasing systems (e.g., Sipper, *Machine Nature*; Forbes, *Imitation of Life*; Amos, *Genesis Machines*).

47. See, for instance, the comparative analysis of sequencing and environmental data flows in McNally et al., "Understanding the 'Intensive' in 'Data Intensive Research.'"

48. Leonelli, "When Humans are the Exception."

49. A clear summary of these arguments is provided by Mayer-Schönberger and Cukier, *Big Data*. A different take on how to tackle the potential unreliability of data resources has been offered by William Wimsatt, in *Re-Engineering Philosophy for Limited Beings*, who, drawing on an analogy between the ways in which organisms cope with malfunctioning and the ways in which computational tools are increasingly engineered to cope with error, has suggested that the robustness of computational infrastructure may be strengthened by mimicking evolutionary and developmental robustness mechanisms.

50. Royal Society, "Science as an Open Enterprise."

51. See Wylie, *Thinking from Things*. One is the opportunity to detect phenomena at a higher level of resolution and accuracy: "To detect smaller effects, we need bigger data." For complex phenomena, we need more data; some types of research simply needs large datasets (e.g., looking for drug compounds in plants, looking for susceptibility genes in cancer research).

52. Within experimental biology, the ability to explain why a certain behavior obtains is still very highly valued—arguably over and above the ability to relate two traits to each other. The value of causal explanations in the life sciences is a key concern for many philosophers, particularly those interested in mechanistic explanations as a form of biological understanding (e.g., Bechtel, *Discovering Cell Mechanisms*; Craver and Darden, *In Search of Biological Mechanisms*).

53. See, for instance, the *Nature* special, "Challenges in Irreproducible Research."

54. Synaptic Leap, "Homepage." See also Guerra et al., "Assembling a Global Database of Malaria Parasite Prevalence for the Malaria Atlas Project."

55. MacLean et al., "Crowdsourcing Genomic Analyses of Ash and Ash Dieback."

56. Kelty, "Outlaws, Hackers, Victorian Amateurs"; Delfanti, *Biohackers*; Wouters et al., *Virtual Knowledge*.

57. Stevens and Richardson, *Postgenomics*.

58. When bringing attention to the diverse and ever-changing nature of research traditions in biology, data-centric methods may well help to harness such diversity to capture the processual dimensions of life, particularly as investigated by epigenetics and evolutionary-developmental biology. Similarly, the willingness to embrace and encourage the distribution of biological reasoning across vast research networks may well foster integrative understandings of organisms in the long term.

CHAPTER SEVEN

1. Azberger et al., "An International Framework to Promote Access to Data"; Organisation for Economic Co-Operation and Development, "OECD Principles and Guidelines for Access to Research Data from Public Funding"; Royal Society, "Science as an Open Enterprise."

2. These quotes are Crombie's original formulation of styles of scientific thinking, which Hacking calls "styles of reasoning" (Crombie, *Styles of Scientific Thinking in the European Tradition*; Hacking, *Historical Ontology*). For a detailed discussion of these notions, see also Kusch, "Hacking's Historical Epistemology."

3. A similar argument can be made with reference to John Pickstone's "ways of knowing," as data centrism involves elements from each of the traditions that he singles out: natural history, analysis, experimentalism, and technoscience (Pickstone, *Ways of knowing*).

4. Lipton, *Inference to the Best Explanation*; Douven, "Abduction."

5. I agree with Woodward that "scientists who are engaged in assessments of [data] reliability do not themselves rely exclusively on logical or formal structural or subject-matter-independent relationships of the sort emphasized in traditional accounts of confirmation in assessing evidential support" ("Data, Phenomena, and Reliability," S171).

6. Going back to my discussion of value in chapter 2, it could be said that in data-centric biology, concerns around data are valued more than concerns around theory.

7. In the case of data journeys, attempting to draw such a distinction is uninformative, if it is at all possible. Each manipulation that data undergo during their travels may be conceptualized both as contexts of discovery, in which researchers shape the ways in which data will ultimately be interpreted and used as evidence for claims, and contexts of justification, since a rational reconstruction of the steps through which data have been interpreted involves tracing and justifying each stage of their journey, which constitutes a difficult task, particularly given the diversity of expertises and motivations involved in making data travel. The convergence of these two functions within the same set of actions, and the challenges confronted by curators and users in making data travel in ways that can be appropriately documented and discussed, is what makes data journeys epistemologically interest-

ing. Both discovery and justification are part and parcel of all the phases of data journeys that I have discussed, and distinguishing between them does not help to understand the shifts in circumstances and significance that data undergo when moving around.

8. See for instance Schickore and Steinle, *Revisiting Discovery and Justification.*

9. For instance, logical empiricists like Carl Hempel and Ernest Nagel dubbed the context of discovery as subjective and irrational and thus beyond the realm of scientific epistemology. The popularity of these views is a prominent reason for the intellectual rift between philosophical and sociological approaches that plagues Anglo-American science studies to this day (see, for instance, Longino, *The Fate of Knowledge*).

10. See, for instance, Ernan McMullin's discussion of epistemic values ("Values in Science") and Rooney's related commentary ("On Values in Science").

11. Rouse, *Knowledge and Power*, 125. See also Gooding, *Experiment and the Making of Meaning*; Franklin, *The Neglect of Experiment*; Collins, *Changing Order*; Latour and Woolgar, *Laboratory Life*; Hacking, *Historical Ontology*; and Radder, *The Material Realization of Science.*

12. Rheinberger, *Towards a History of Epistemic Things* and *An Epistemology of the Concrete.*

13. Chang, *Is Water H₂O?*, 16.

14. For instance, Martin Kusch has criticized Chang's work for its lack of detailed engagement with social and political circumstances of the Chemical Revolution, arguing that this led Chang to unwittingly equate the "social" with the "irrational" (Kusch, "Scientific Pluralism and the Chemical Revolution").

15. This is particularly evident in Rheinberger's recent *Cultural History of Heredity*, coauthored with Staffan Müller-Wille, where a clear attempt is made to intertwine a broad social and cultural analysis with a technical study of the origins and development of genetics.

16. "Scientific knowledge must be understood in its use. This use involves a local, existential knowledge located in a circumspective grasp of the configuration of institutions, social roles, equipment, and practices that make science an intelligible activity in our world" (Rouse, *Knowledge and Power*, 126).

17. For pioneering philosophical work along these lines, see Douglas, *Science, Policy, and the Value-Free Ideal*; Elliott, *Is a Little Pollution Good for You?*; and Biddle, "Can Patents Prohibit Research?" Drawing such connections was the explicit goal of the scholars involved in the Knowledge/Value workshop series coordinated by Kaushik Sunder Rajan, in which I was privileged to take part between 2009 and 2014 (Knowledge/Value website).

18. Even among philosophers who devote attention to the meaning of context in their account, we find widely diverging interpretations. For instance, Sandra Mitchell interprets context-dependence as applying to biological phenomena and thus to "the very fabric of the world" (*Unsimple Truths*, 65). Peter Vickers uses it instead in reference to argumentative practices, where "that one and the same argument may be sound in one context and not in another"—an interpretation that makes context-dependence "entail relativity and loss of objectivity" (Vickers, "The Problem of Induction"). In yet another interpretation, which comes closer to my own approach, Paul Hoyningen-Huene uses it to describe the local manifesta-

NOTES TO PAGES 181–184

tions of systematicity in scientific knowledge, which are always grounded in specific circumstances, including the frameworks and goals of the field in which knowledge is pursued (*Systematicity*, 26).

19. Longino, *The Fate of Knowledge*, 1. See also her *Science as Social Knowledge*.

20. It is no accident that Longino uses examples from data practices as her own starting point to investigate the social dimensions of research. Like me, she views underdetermination as a starting point for scientific inquiry rather than an obstacle to logical reasoning: "the consequence of the gap between evidence and hypotheses" (*The Fate of Knowledge*, 127), which can only be filled by "treating agents or subjects of knowledge as located in particular and complex interrelationships and acknowledging that purely logical constraints cannot compel them to accept a particular theory" (*The Fate of Knowledge*, 127–28).

21. My reading of Reichenbach is informed by Don Howard's "Lost Wanderers in the Forest of Knowledge," a study of Reichenbach's motivations in writing *Experience and Prediction* and by Ken Waters's discussion in "What Concept Analysis in Philosophy Should Be" of Reichenbach's views on conceptual analysis and its goals. As Waters emphasizes, Reichenbach starts his book by acknowledging that "knowledge, therefore, is a very concrete thing; and the examination into its properties means studying the features of a sociological phenomenon" (Reichenbach, *Experience and Prediction*, 3).

22. Reichenbach, *Experience and Prediction*, 3.

23. Dewey grounds this conceptualization of research on the "principle of the continuum of inquiry" (*Logic*, 2). My reading of Dewey's logic, particularly of his notion of situation, is informed by Matthew J. Brown's detailed discussion of its relation to the philosophy of science in "John Dewey's Logic of Science."

24. Dewey, *Logic*, 68.

25. Ibid.

26. Ibid, 104. I am grateful to Erik Weber for exhorting me to emphasize this crucial aspect.

27. "Facts are evidential and are tests of an idea in so far as they are capable of being organized with one another" (Dewey, *Logic*, 113).

28. I already discussed the extensive consultations underlying the development of iPlant and the difficulties encountered by caBIG in its strategy for data dissemination. See Waters, "How Practical Know-How Contextualizes Theoretical Knowledge," for a discussion of the conceptual implications of the lack of technologies to order and visualize genetic data in the early twentieth century.

29. In such cases, objects that at one time were viewed as data become waste, since it is impossible to ascribe evidential value to them (see chapter 3). Given the vast amounts of data generated by scientists, this is an unavoidable part of the research cycle, and indeed the vast majority of scientific data are eventually lost or forgotten (Bowker, *Memory Practices in the Sciences*). The crucial question within data-centric research is whether there can be any systematic, reasoned approach to which data are preserved and which are lost, and why—following which, researchers would be able to decide and justify which of their data they preserve and in which form.

30. Sheila Jasanoff originally introduced the notion of coproduction in a

broader setting, to discuss the relation between science and social order (*States of Knowledge*); Stephen Hilgartner has usefully applied it to the constitution of property regimes within biological laboratories (Hilgartner, "Mapping Systems and Moral Order").

31. Russell, *The Basic Writings of Bertrand Russell*, 180.

32. Brown, "John Dewey's Logic of Science."

33. This is comparable to what Wimsatt and Griesemer call "entrenchment" in "Reproducing Entrenchments to Scaffold Culture." See also Leonelli and Ankeny, "Repertoires."

34. Note that I am not describing the context of data production as one where considerations about data dissemination and interpretation do not matter, which would be patently false in the vast majority of cases (at least within biology); rather, I am arguing that conditions of production are researchers' *primary* concern here.

35. It is worth noting here that the increasing use of doctor consultations as means of data gathering via centralized databases may well change the role of physicians in mediating information between patients and general repositories. In order for physicians to be true to their oath of respect for patient confidentiality, they will need to pay attention not only to their own situation of data production but also to the subsequent journeys of data and the extent to which they may become mobilized in ways that were not intended at the time of the consultation.

36. See, for instance, Popper, "Models, Instruments, and Truth," on "situational analysis," Forrester, "If *p*, Then What?"; Radder, *The World Observed / The World Conceived*; and Morgan, *The World in the Model* (ch. 9 on "typical situations").

37. Ankeny, "The Overlooked Role of Cases in Causal Attribution in Medicine"; see also Ankeny, "Detecting Themes and Variations."

38. Building on an extensive study of case-based reasoning across disciplines, Morgan provided a taxonomy of strategies through which knowledge is resituated (Morgan, "Resituating Knowledge: Generic Strategies and Case Studies").

39. This suggestion could be expanded to think of data epistemology as primarily an epistemology of cases rather than of all-encompassing Kuhnian paradigms or widely applicable theoretical frameworks. Each situation of data handling is both unique and potentially continuous with others. When continuities are strong (often as a result of scientific standardization and control mechanisms in research), they can be mistaken for generalizations about supposedly incommensurable, coherent paradigms; but the hold that specific situations have on researchers' imagination and ways of acting is contingent on material and social factors that can change at any time. This position thus builds on Kuhn, by acknowledging importance of material and social conditions of knowledge production, but also supersedes Kuhnian views on theory change by rejecting his attempt to arrange situations of inquiry into sweepingly broad and self-contained trends.

40. Steve Hinchliffe has introduced the idea of "knowing around" organisms as a way to challenge linear perceptions of "the range of practices, materials, and movements involved in making life knowable" (Hinchliffe and Lavau, "Differentiated Circuits," 259). I find this expression congenial to the study of data: what databases enable users to do in my case is to "know around" data, in the sense

of learning to contextualize them in ways that help their understanding of their potential biological significance.

41. Joseph Rouse, in *Knowledge and Power*, has made a similar point when discussing the travel of knowledge outside its original context of production: "Knowledge is extended outside the laboratory not by generalisation to universal laws instantiable elsewhere, but by the adaptation of locally situated practices to new local contexts." In my framework, I am extending this analysis to encompass a general reflection on processes of contextualization and their importance with specific regard to data travels.

42. Haraway, "Situated Knowledges," 590.

43. For an expanded argument about how the notion of situatedness can be applied to experimental settings, see Leonelli et al., "Making Organisms Model Humans."

44. Christine Hine has highlighted the dissonance among accountabilities involved in data handling decisions, which often leads the researchers involved to juggle a variety of nonoverlapping (and sometimes contradictory) commit-ments—a process that Hine calls "dance of initiatives" (Hine, "Databases as Scientific Instruments and Their Role in the Ordering of Scientific Work").

45. For details on ATLAS data dissemination, see the CERN website, http://www.cern.ch; for details on ATLAS data-selection procedures, see Karaca, "A Study in the Philosophy of Experimental Exploration."

Bibliography

Agar, Jon. "What Difference Did Computers Make?" *Social Studies of Science* 36(6) (2006): 869–907.

Allen, John F. "Bioinformatics and Discovery: Induction Beckons Again." *BioEssays* 23(1) (2001): 104–7.

Amos, Martyn. *Genesis Machines: The New Science of Biocomputation*. London: Atlantic Books, 2014.

Anderson, Chris. "The End of Theory: The Data Deluge Makes the Scientific Method Obsolete." *Wired Magazine*, June 23, 2008. Accessed October 17, 2015. http://www.wired.com/science/discoveries/magazine/16-07/pb_theory.

Ankeny, Rachel A. "Detecting Themes and Variations: The Use of Cases in Developmental Biology." *Philosophy of Science* 79(5) (2012): 644–54.

———. "The Natural History of *Caenorhabditis elegans* Research." *Nature Reviews Genetics* 2 (2001): 474–79.

———. "The Overlooked Role of Cases in Causal Attribution in Medicine." *Philosophy of Science* 81 (2014): 999–1011.

Ankeny, Rachel A., Hasok Chang, Marcel Boumans, and Mieke Boon. "Introduction: Philosophy of Science in Practice." *European Journal for Philosophy of Science* 1(3) (2011): 303–7.

Ankeny, Rachel A., and Sabina Leonelli. "Valuing Data in Postgenomic Biology: How Data Donation and Curation Practices Challenge the Scientific Publication System." In *Postgenomics*, edited by Sarah Richardson and Hallam Stevens, 126–49. Durham, NC: Duke University Press, 2015.

———. "What's So Special about Model Organisms?" *Studies in History and Philosophy of Science* 42(2) (2011): 313–23.

Aronova, Elena. "Environmental Monitoring in the Making: From

Surveying Nature's Resources to Monitoring Nature's Change." *Historical Social Research* 40(2) (2015): 222–45.

Aronova, Elena, Karen Baker, and Naomi Oreskes. "Big Science and Big Data in Biology: From the International Geophysical Year through the International Biological Program to the Long Term Ecological Research (LTER) Network, 1957–Present." *Historical Studies in the Natural Sciences* 40(2) (2010): 183–224.

Ashburner, Michael, Catherine A. Ball, Judith A. Blake, David Botstein, Heather Butler, J. Michael Cherry, Allan P. Davis et al. "Gene Ontology: Tool for the Unification of Biology." *Nature Genetics* 25 (2000): 25–29.

Augen, Jeff. *Bioinformatics in the Post-Genomic Era: Genome, Transcriptome, Proteome, and Information-Based Medicine.* Boston: Addison-Wesley, 2005.

Avraham, Shulamit, Chih-Wei Tung, Katica Ilic, Pankaj Jaiswal, Elizabeth A. Kellogg, Susan McCouch, Anuradha Pujar et al. "The Plant Ontology Database: A Community Resource for Plant Structure and Developmental Stages Controlled Vocabulary and Annotations." *Nucleic Acids Research* 36(S1) (2008): D449–D454.

Azberger, Peter, Peter Schroeder, Anne Beaulieu, Geoffrey Bowker, Kathleen Casey, Leif Laaksonen, David Moorman et al. "An International Framework to Promote Access to Data." *Science* 303(5665) (2004): 1777–78.

Babcock, Bruce A. *The Impact of US Biofuel Policies on Agricultural Price Levels and Volatility.* International Centre for Trade and Sustainable Development Issue Paper No. 35. Geneva, Switzerland: ICTSD, 2011. Accessed October 17, 2015. http://www.ictsd.org/downloads/2011/12/the-impact-of-us-biofuel-policies-on -agricultural-price-levels-and-volatility.pdf. Republished in *China Agricultural Economic Review* 4(4) (2012): 407–26.

Baclawski, Kenneth, and Tianhua Niu. *Ontologies for Bioinformatics.* Cambridge, MA: MIT Press, 2006.

Bacon, Francis. *The Novum Organum: With Other Parts of the Great Instauration.* Edited and translated by Peter Urbach and John Gibson. La Salle, IL: Open Court, 1994.

Bada, Michael, Robert Stevens, Carole Goble, Yolanda Gil, Michael Ashburner, Judith A. Blake, J. Michael Cherry et al. "A Short Study on the Success of the Gene Ontology." *Journal of Web Semantics* 1(2) (2004): 38.

Bafoutsou, Georgia, and Gregoris Mentzas. "Review and Functional Classification of Collaborative Systems." *International Journal of Information Management* 22(4) (2002): 281–305.

Bailer-Jones, Daniela M. *Scientific Models in Philosophy of Science.* Pittsburgh, PA: University of Pittsburgh, 2009.

Baker, Karen S., and Florence Millerand. "Infrastructuring Ecology: Challenges in Achieving Data Sharing." In *Collaboration in the New Life Sciences*, edited by John N. Parker, Niki Vermeulen, and Bart Penders, 111–38. Farnham, UK: Ashgate, 2010.

Bard, Jonathan B. L., and Seung Y. Rhee. "Ontologies in Biology: Design, Applications and Future Challenges." *Nature Reviews Genetics* 5 (2004): 213–22.

Barnes, Barry, and John Dupré. *Genomes and What to Make of Them.* Chicago: University of Chicago Press, 2008.

Barry, Andrew. *Political Machines: Governing a Technological Society*. London: Athlone Press, 2001.

Bastow, Ruth, Jim Beynon, Mark Estelle, Joanna Friesner, Erich Grotewold, Irene Lavagi, Keith Lindsey et al. "An International Bioinformatics Infrastructure to Underpin the Arabidopsis Community." *The Plant Cell* 22 (2010): 2530–36.

Bastow, Ruth, and Sabina Leonelli. "Sustainable Digital Infrastructure." *EMBO Reports* 11(10) (2010): 730–35.

Bauer, Susanne. "Mining Data, Gathering Variables, and Recombining Information: The Flexible Architecture of Epidemiological Studies." *Studies in History and Philosophy of Biological and Biomedical Sciences* 39 (2008): 415–26.

Bechtel, William. *Discovering Cell Mechanisms: The Creation of Modern Cell Biology*. Cambridge: Cambridge University Press, 2006.

———. "From Molecules to Behavior and the Clinic: Integration in Chronobiology." *Studies in History and Philosophy of Biological and Biomedical Sciences* 44(4) (2013): 493–502.

Bechtel, William, and Robert C. Richardson. *Discovering Complexity: Decomposition and Localization as Strategies in Scientific Research*. Princeton, NJ: Princeton University Press, 1993.

Behringer, Richard R., Alexander D. Johnson, and Robert E. Krumlauf, eds. *Emerging Model Organisms: A Laboratory Manual, Volume 1*. Cold Spring Harbor, NY: Cold Spring Harbor Press, 2008.

———, eds. *Emerging Model Organisms: A Laboratory Manual, Volume 2*. Cold Spring Harbor, NY: Cold Spring Harbor Press, 2008.

Bernard, Claude. *An Introduction to the Study of Experimental Medicine*. Mineola, NY: Dover, [1855] 1957.

Bevan, Michael, and Sean Walsh. "Positioning Arabidopsis in Plant Biology. A Key Step Toward Unification of Plant Research." *Plant Physiology* 135 (2004): 602–6.

Bezuidenhout, Louise, Brian Rappert, Ann Kelly, and Sabina Leonelli. "Open Data and the Digital Divide in Life Science Research." Forthcoming.

Biagioli, Mario. "Plagiarism, Kinship and Slavery." *Theory, Culture and Society* 31(2–3) (2014): 65–91. doi:10.1177/0263276413516372.

Biddle, Justin. "Can Patents Prohibit Research? On the Social Epistemology of Patenting and Licensing in Science." *Studies in History and Philosophy of Science* 45 (2014): 14–23.

BioSharing. "Homepage." Accessed April 30, 2014. https://biosharing.org.

Blair, Ann. *Too Much to Know: Managing Scholarly Information before the Modern Age*. New Haven, CT: Yale University Press, 2010.

Blount, Zachary D., Jeffrey E. Barrick, Carla J. Davidson, and Richard E. Lenski. "Genomic Analysis of a Key Innovation in an Experimental *Escherichia coli* Population." *Nature* 489 (2012): 513–18.

Bogen, James. "Noise in the World." *Philosophy of Science* 77 (2010): 778–91.

———. "Theory and Observation in Science." In *The Stanford Encyclopedia of Philosophy*, edited by Edward N. Zalta et al. Accessed October 17, 2015. http://plato.stanford.edu/entries/science-theory-observation.

Bogen, James, and James Woodward. "Saving the Phenomena." *The Philosophical Review* 97(3) (1988): 303–52.

Bolker, Jessica A. "Model Organisms: There's More to Life than Rats and Flies." *Nature* 491 (2012): 31–33.

———. "Model Systems in Developmental Biology." *BioEssays* 17 (1995): 451–55.

Borgman, Christine L. "The Conundrum of Sharing Research Data." *Journal of the American Society for Information Science and Technology* 63(6) (2012): 1059–78.

———. *Scholarship in the Digital Age: Information, Infrastructure, and the Internet.* Cambridge, MA: MIT Press, 2007.

Botanical Society of America. "Botany for the Next Millennium: I. The Intellectual: Evolution, Development, Ecosystems." 1994. Accessed October 17, 2015. http://www.botany.org/bsa/millen/mil-chp1.html#Evolution.

Boumans, Marcel, and Sabina Leonelli. "Introduction: On the Philosophy of Science in Practice." *Journal for General Philosophy of Science / Zeitschrift für Allgemeine Wissenschaftstheorie* 44(2) (2013): 259–61.

Bowker, Geoffrey C. "Biodiversity Datadiversity." *Social Studies of Science* 30(5) (2000): 643–83.

———. *Memory Practices in the Sciences.* Cambridge, MA: MIT Press, 2006.

Bowker, Geoffrey C., and Susan Leigh Star. *Sorting Things Out: Classification and Its Consequences.* Cambridge, MA: MIT Press, 1999.

Brazma, Alvis, et al. "Minimal Information Around a Microarray Experiment (MIAME): Towards Standards for Microarray Data." Nature Genetics 29(4) (2001): 365–71.

Brazma, Alvis, Maria Krestyaninova, and Ugis Sarkans. "Standards for Systems Biology." *Nature Reviews Genetics* 7 (2006): 593–605.

Brazma, Alvis, Alan Robinson, and Jaak Vilo. "Gene Expression Data Mining and Analysis." In *DNA Microarrays: Gene Expression Applications*, edited by Bertrand Jordan, 106–32. Berlin: Springer, 2001.

Brigandt, Ingo. "Beyond Reduction and Pluralism: Toward an Epistemology of Explanatory Integration in Biology." *Erkenntnis* 73(3) (2010): 295–311.

———. "Natural Kinds in Evolution and Systematics: Metaphysical and Epistemological Considerations." *Acta Biotheoretica* 57 (2009): 77–97.

———, ed. "Special Section—Integration in Biology: Philosophical Perspectives on the Dynamics of Interdisciplinarity." *Studies in History and Philosophy of Biological and Biomedical Sciences* 4(A) (2013): 461–571.

———. "Systems Biology and the Integration of Mechanistic Explanation and Mathematical Explanation." *Studies in History and Philosophy of Biological and Biomedical Sciences* 44(4) (2013): 477–92.

Brown, Matthew J. "John Dewey's Logic of Science." *HOPOS: The Journal of the International Society for the History of Philosophy of Science* 2(2) (2012): 258–306.

Browne, Janet. "History of Plant Sciences." *eLS.* March 16, 2015. Accessed October 17, 2015. http://onlinelibrary.wiley.com/doi/10.1002/9780470015902.a0003081.pub2/abstract. doi:10.1002/9780470015902.a0003081.pub2.

Bülow, Lorenz, Martin Schindler, and Reinhard Hehl. "PathoPlant®: A Platform for Microarray Expression Data to Analyze Co-Regulated Genes Involved in Plant Defense Responses." *Nucleic Acids Research* 35 (2007): 841–45.

Bult, Carol J. "From Information to Understanding: the Role of Model Organism Databases in Comparative and Functional Genomics." *Animal Genetics* 37(S1) (2006): 28–40.

Burian, Richard M. "The Dilemma of Case Studies Resolved: The Virtues of Using Case Studies in the History and Philosophy of Science." *Perspectives on Science* 9 (2001): 383–404.

———. "Exploratory Experimentation and the Role of Histochemical Techniques in the Work of Jean Brachet, 1938–1952." *History and Philosophy of the Life Sciences* 19 (1997): 27–45.

———. "How the Choice of Experimental Organism Matters: Epistemological Reflections on an Aspect of Biological Practice." *Journal of the History of Biology* 26(2) (1993): 351–67.

———. "More than a Marriage of Convenience: On the Inextricability of History and Philosophy of Science." *Philosophy of Science* 44 (1977): 1–42.

———. "On microRNA and the Need for Exploratory Experimentation in Post-Genomic Molecular Biology." *History and Philosophy of the Life Sciences* 29(3) (2007): 285–312.

Callebaut, Werner. "Scientific Perspectivism: A Philosopher of Science's Response to the Challenge of Big Data Biology." *Studies in History and Philosophy of Biological and Biomedical Sciences* 43(1) (2012): 69–80.

Cambrosio, Alberto, Peter Keating, Thomas Schlich, and George Weisz. "Regulatory Objectivity and the Generation and Management of Evidence in Medicine." *Social Science and Medicine* 63(1) (2006): 189–99.

Caporael, Linnda R., James R. Griesemer, and William C. Wimsatt, eds. *Developing Scaffolds in Evolution, Culture, and Cognition.* Cambridge, MA: MIT Press, 2013.

Cartwright, Nancy. *The Dappled World: A Study of the Boundaries of Science.* Cambridge: Cambridge University Press, 2002.

———. *How the Laws of Physics Lie.* Oxford: Oxford University Press, 1983.

Casadevall, Arturo, and Ferric C. Fang. "Reproducible Science." *Infection and Immunity* 78(12) (2010): 4972–75.

Chang, Hasok. *Inventing Temperature: Measurement and Scientific Progress.* New York: Oxford University Press, 2004.

———. *Is Water H_2O?: Evidence, Realism and Pluralism.* Dordrecht, Netherlands: Springer, 2012.

Chisholm, Rex L., Pascale Gaudet, Eric M. Just, Karen E. Pilcher, Petra Fey, Sohel N. Merchant, and Warren A. Kibbe. "dictyBase, the Model Organism Database for *Dictyostelium discoideum.*" *Nucleic Acids Research* 34 (2006): D423–D427.

Chow-White, Peter A., and Miguel García-Sancho. "Bidirectional Shaping and Spaces of Convergence: Interactions between Biology and Computing from the First DNA Sequencers to Global Genome Databases." *Science, Technology, and Human Values* 37(1) (2012): 124–64.

Clark, Andy. *Being There: Putting Brain, Body, and World Together Again.* Cambridge, MA: MIT Press, 1997.

Clarke, Adele E., and Joan H. Fujimura, eds. *The Right Tools for the Job: At Work in Twentieth-Century Life Sciences.* Princeton, NJ: Princeton University Press, 1992.

Collins, Harry M. *Changing Order: Replication and Induction in Scientific Practice.* Chicago: University of Chicago Press, 1985.

Computational Biology Research Group of the University of Oxford. "Examples of Common Sequence File Formats." Accessed February 2014. http://www.comp bio.ox.ac.uk/bioinformatics_faq/format_examples.shtml.

Cook-Deegan, Robert. *The Gene Wars: Science, Politics and the Human Genome.* New York: W. W. Norton, 1994.

————. "The Science Commons in Health Research: Structure, Function, and Value." *Journal of Technology Transfer* 32 (2007): 133–56.

Craver, Carl F. "Beyond Reduction: Mechanisms, Multifield Integration and the Unity of Neuroscience." *Studies in History and Philosophy of Biological and Biomedical Sciences* 36(2) (2005): 373–95.

Craver, Carl F., and Lindley Darden. *In Search of Biological Mechanisms: Discoveries Across the Life Sciences.* Chicago: University of Chicago Press, 2013.

Creager, Angela N. H., Elizabeth Lunbeck, and M. Norton Wise. *Science without Laws: Model Systems, Cases, Exemplary Narratives.* Durham, NC: Duke University Press, 2007.

Crombie, Alistair C. *Styles of Scientific Thinking in the European Tradition: The History of Argument and Explanation Especially in the Mathematical and Biomedical Sciences and Arts.* London: Gerald Duckworth, 1994.

Cukier, Kenneth. "Data, Data Everywhere." *The Economist*, February 25, 2010. Accessed October 17, 2015. http://www.economist.com/node/15557443.

Darden, Lindley. "Relations Among Fields: Mendelian, Cytological and Molecular Mechanisms." *Studies in History and Philosophy of Biological and Biomedical Sciences* 36(2) (2005): 349–71.

Darden, Lindley, and Nancy Maull. "Interfield Theories." *Philosophy of Science* 44(1) (1977): 43–64.

Daston, Lorraine. "Type Specimens and Scientific Memory." *Critical Inquiry* 31(1) (2004): 153–82.

Daston, Lorraine, and Peter Galison. "The Image of Objectivity." *Representations* 40 (1992): 81–128.

Daston, Lorraine, and Elisabeth Lunbeck. *Histories of Scientific Observation.* Chicago: University of Chicago Press, 2010.

Davies, Gail. "Arguably Big Biology: Sociology, Spatiality and the Knockout Mouse Project." *BioSocieties* 4(8) (2013): 417–31.

————. "What Is a Humanized Mouse? Remaking the Species and Spaces of Translational Medicine." *Body and Society* 18 (2012): 126–55.

Davies, Gail, Emma Frow, and Sabina Leonelli. "Bigger, Faster, Better? Rhetorics and Practices of Large-Scale Research in Contemporary Bioscience." *BioSocieties* 8(4) (2013): 386–96.

de Chadarevian, Soraya. *Designs for Life: Molecular Biology after World War II.* Cambridge: Cambridge University Press, 2002.

————. "Of Worms and Programmes: Caenorhabditis elegans and the Study of Development." *Studies in History and Philosophy of Biology and Biomedical Sciences* 1 (1998): 81–105.

de Chadarevian, Soraya, and Nick Hopwood, eds. *Models: The Third Dimension of Science.* Stanford, CA: Stanford University Press, 2004.

De Luca, Vincenzo, Vonny Salim, Sayaka Masada Atsumi, and Fang Yu. "Mining the Biodiversity of Plants: A Revolution in the Making." *Science* 336(6089) (2012): 1658–61.

Delfanti, Alessandro. *Biohackers: The Politics of Open Science.* London: Pluto Press, 2013.

de Regt, Henk W., Sabina Leonelli, and Kai Eigner, eds. *Scientific Understanding: Philosophical Perspectives*. Pittsburgh, PA: University of Pittsburgh Press, 2009.

Dewey, John. *Logic: The Theory of Inquiry*. New York: Holt, Rinehart, and Winston, 1938.

———. "The Need for a Recovery in Philosophy." In *The Pragmatism Reader: From Peirce Through the Present*, edited by Robert B. Talisse and Scott F. Aikin. Princeton, NJ: Princeton University Press, 2011.

Douglas, Heather E. *Science, Policy, and the Value-Free Ideal*. Pittsburgh, PA: University of Pittsburgh Press, 2009.

Douven, Igor. "Abduction." In *The Stanford Encyclopedia of Philosophy*, edited by Edward N. Zalta et al. Accessed October 17, 2015. http://plato.stanford.edu/entries/abduction.

Draghici, Sorin, Puvesh Khatri, Aron C. Eklund, and Zoltan Szallasi. "Reliability and Reproducibility Issues in DNA Microarray Measurements." *Trends in Genetics* 22(2) (2006): 101–19.

Dupré, John. *The Disorder of Things*. Cambridge: Cambridge University Press, 1993.

———. "In Defence of Classification." *Studies in History and Philosophy of Biological and Biomedical Sciences* 32 (2001): 203–19.

———. *Processes of Life*. Oxford: Oxford University Press, 2012.

Dupré, John, and Maureen A. O'Malley. "Metagenomics and Biological Ontology." *Studies in History and Philosophy of Biological and Biomedical Sciences* 38 (2007): 834–46.

Dussauge, Isabelle, Claes-Fredrik Helgesson, and Francis Lee. *Value Practices in the Life Sciences and Medicine*. Oxford: Oxford University Press, 2015.

Edwards, Paul N. *A Vast Machine: Computer Models, Climate Data, and the Politics of Global Warming*. Cambridge, MA: MIT Press, 2010.

Edwards, Paul N., Matthew S. Mayernik, Archer L. Batcheller, Geoffrey C. Bowker, and Christine L. Borgman. "Science Friction: Data, Metadata, and Collaboration." *Social Studies of Science* 41(5) (2011): 667–90.

Egaña Aranguren, Mikel, Sean Bechhofer, Phillip Lord, Ulrike Sattler, and Robert Stevens. "Understanding and Using the Meaning of Statements in a Bio-Ontology: Recasting the Gene Ontology in OWL." *BMC Bioinformatics* 8 (2007): 57.

Elixir. "Elixir: A Distributed Infrastructure for Life Sciences Information." Accessed October 17, 2015. http://www.elixir-europe.org.

Elliott, Kevin C. *Is a Little Pollution Good for You? Incorporating Societal Values in Environmental Research*. New York: Oxford University Press, 2011.

Endersby, Jim. *A Guinea Pig's History of Biology*. Cambridge, MA: Harvard University Press, 2007.

———. *Imperial Nature: Joseph Hooker and the Practices of Victorian Science*. Chicago: University of Chicago Press, 2008.

European Bioinformatics Institute. "Malaria Data." Accessed October 17, 2015. https://www.ebi.ac.uk/chembl/malaria.

Evans, James, and Andrey Rzhetsky. "Machine Science." *Science* 329(5990) (2010): 399–400.

Fecher, Benedikt, Sascha Friesike, and Marcel Hebing. "What Drives Academic Data Sharing?" *PLOS ONE* 10(2) (2015): e0118053. doi:10.1371/journal.pone.0118053.

Fehr, Carla, and Katherine Plaisance, eds. "Making Philosophy of Science More Socially Relevant." Special issue, *Synthese* 177(3) (2010).

Fernie, Alisdair R., Asaph Aharoni, Lothar Wilmitzer, Mark Stitt, Takayuki Tohge,
Joachim Kopta, Adam J. Carroll et al. "Recommendations for Reporting Metabolite Data." *Plant Cell* 23 (2011): 2477–82.

Finch Group. "Accessibility, Sustainability, Excellence: How to Expand Access to
Research Publications." 2012 Report of the Working Group on Expanding Access to Published Research Findings. Accessed October 17, 2015. http://www
.researchinfonet.org/wp-content/uploads/2012/06/Finch-Group-report-FINAL
-VERSION.pdf.

Finholt, Thomas A. "Collaboratories." *Annual Review of Information Science and
Technology* 36 (2002): 73–107.

Floridi, Luciano. "Is Information Meaningful Data?" *Philosophy and Phenomeno-
logical Research* 70(2) (2005): 351–70.

———. *The Philosophy of Information.* Oxford: Oxford University Press, 2011.

———. "Semantic Conceptions of Information." In *The Stanford Encyclopedia of
Philosophy*, edited by Edward N. Zalta et al. Accessed October 17, 2015. http://
plato.stanford.edu/entries/information-semantic.

Floridi, Luciano, and Phyllis Illari. *The Philosophy of Information Quality.* Synthese
Library 358. Cham, Switzerland: Springer, 2014.

The FlyBase Consortium. "The FlyBase Database of the Drosophila Genome Projects and Community Literature." *Nucleic Acids Research* 30 (2002): 106–8.

Forbes, Nancy. *Imitation of Life: How Biology Is Inspiring Computing.* Cambridge,
MA: MIT Press, 2005.

Forrester, John. "If *p*, Then What? Thinking in Cases." *History of the Human
Sciences* 9(3) (1996): 1–25.

Fortun, Michael. "The Care of the Data." Accessed October 17, 2015. http://mfortun
.org/?page_id=72.

———. *Promising Genomics: Iceland and deCODE Genetics in a World of Specula-
tion.* Berkeley, CA: University of California Press, 2008.

Foucault, Michel. *The Birth of the Clinic.* London: Routledge, 1973.

Franklin, Allan. *The Neglect of Experiment.* Cambridge: Cambridge University
Press, 1986.

Friese, Carrie, and Adele E. Clarke. "Transposing Bodies of Knowledge and Technique: Animal Models at Work in Reproductive Sciences." *Social Studies of
Science* 42 (2011): 31–52.

Frigg, Roman, and Stephen Hartmann. "Models in Science." In *The Stanford Encyclopedia of Philosophy*, edited by Edward N. Zalta et al. Accessed October 17,
2015. http://plato.stanford.edu/entries/models-science.

Fry, Ben. *Visualizing Data: Exploring and Explaining Data with the Processing Environment.* Sebastopol, CA: O'Reilly Media, 2007.

Galison, Peter. *How Experiments End.* Chicago: University of Chicago Press, 1987.

———. *Image and Logic: A Material Culture of Microphysics.* Chicago: University
of Chicago Press, 1997.

Garcia-Hernandez, Margarita, Tanya Z. Berardini, Guanghong Chen, Debbie Crist,
Aisling Doyle, Eva Huala, Emma Knee et al. "TAIR: A Resource for Integrated
Arabidopsis Data." *Functional and Integrative Genomics* 2 (2002): 239–53.

García-Sancho, Miguel. *Biology, Computing, and the History of Molecular Sequencing: From Proteins to DNA, 1945–2000.* New York: Palgrave Macmillan, 2012.

Gene Ontology Consortium. "AmiGO 2," set of tools for searching and browsing the Gene Ontology database. Accessed October 17, 2015. http://geneontology.org.

———. "The Gene Ontology Consortium: Going Forward." *Nucleic Acids Research* 43 (2015): D1049–D1056.

———. "GO:0046031." Accessed October 17, 2015. http://amigo.geneontology.org /amigo/term/GO:0046031.

———. "Minutes of the Gene Ontology Consortium Meeting. (2006)." Taken at St. Croix, US Virgin Islands, March 31–April 2, 2006. Accessed March 12, 2014. http://www.geneontology.org/minutes/20060331_StCroix.pdf.

———. "Minutes of the Gene Ontology Consortium Meeting. (2007)." Taken at Jesus College, Cambridge, UK, January 8–10, 2007. Accessed October 17, 2015. http://www.geneontology.org/meeting/minutes/20070108_Cambridge.doc.

———. "Minutes of the Gene Ontology Content Meeting. (2004)." Taken at Stanford University, California, August 22–23, 2004. Accessed March 12, 2014. http:// www.geneontology.org/GO.meetings.shtml#cont.

Ghiselin, Michael T. "A Radical Solution to the Species Problem" *Systematic Zoology* 23 (1974): 534–46.

Gibbons, Susan M. "From Principles to Practice: Implementing Genetic Database Governance." *Medical Law International* 9(2) (2008): 101–9.

Giere, Ronald N. *Scientific Perspectivism.* Chicago: University of Chicago Press, 2006.

———. *Science without Laws.* Chicago: University of Chicago Press, 1999.

Gitelman, Lisa. *"Raw Data" Is an Oxymoron.* Cambridge, MA: MIT Press, 2013.

GlaxoSmithKline. "Data Transparency." Accessed October 17, 2015. http://www .gsk.com/en-gb/our-stories/how-we-do-randd/data-transparency.

Goble, Carole, and Chris Wroe. "The Montagues and the Capulets." *Comparative and Functional Genomics* 8 (2004): 623–32.

Goff, Stephen A., Matthew Vaughn, Sheldon McKay, Eric Lyons, Ann E. Stapleton, Damian Gessler, Naim Matasci et al. "The iPlant Collaborative: Cyberinfrastructure for Plant Biology." *Frontiers in Plant Science* 2 (2011): 34.

Gooding, David C. *Experiment and the Making of Meaning.* Dordrecht, Netherlands: Kluwer, 1990.

Griesemer, James R. "Formalization and the Meaning of 'Theory' in the Inexact Biological Sciences." *Biological Theory* 7(4) (2013): 298–310.

———. "Periodization and Models in Historical Biology." In *New Perspectives on the History of Life: Essays on Systematic Biology as Historical Narrative*, edited by Michael T. Ghiselin and Giovanni Pinna, 19–30. San Francisco, CA: California Academy of Sciences, 1996.

———. "Reproduction and the Reduction of Genetics." In *The Concept of the Gene in Development and Evolution: Historical and Epistemological Perspectives*, edited by Peter J. Beurton, Raphael Falk, and Hans-Jörg Rheinberger, 240–85. Cambridge: Cambridge University Press, 2000.

———. "Theoretical Integration, Cooperation, and Theories as Tracking Devices." *Biological Theory* 1(1) (2006): 4–7.

Griffiths, Paul E., and Karola Stotz. *Genetics and Philosophy: An Introduction.* Cambridge: Cambridge University Press, 2013.

Gruber, Thomas R. "A Translation Approach to Portable Ontology Specifications." *Knowledge Acquisition* 5(2) (1993): 199–220.

Grundmann, Reiner. "'Climategate' and the Scientific Ethos." *Science, Technology, and Human Values* 38(1) (2012): 67–93.

Guay, Alexandre, and Thomas Pradeau. *Individuals Across the Sciences.* New York: Oxford University Press, 2015.

Guerra, Carlos A., Simon I. Hay, Lorena S. Lucioparedes, Priscilla W. Gikandi, Andrew J. Tatem, Abdisalan M. Noor, and Robert W. Snow. "Assembling a Global Database of Malaria Parasite Prevalence for the Malaria Atlas Project." *Malaria Journal* 6 (2007): 17.

Hacking, Ian. *Historical Ontology.* Cambridge, MA: Harvard University Press, 2002.

———. *Representing and Intervening: Introductory Topics in the Philosophy of Natural Science.* Cambridge: Cambridge University Press, 1983.

———. "The Self-Vindication of the Laboratory Sciences." In *Science as Practice and Culture*, edited by Andrew Pickering, 29–64. Chicago: University of Chicago Press, 1992.

Hanson, Norwood R. *Patterns of Discovery.* Cambridge: Cambridge University Press, 1958.

Haraway, Donna. "Situated Knowledges: The Science Question in Feminism and the Privilege of Partial Perspective." *Feminist Studies* 14(3) (1988): 575–99.

Harris, Todd W., Raymond Lee, Erich Schwarz, Keith Bradnam, Daniel Lawson, Wen Chen, Darin Blasier et al. "WormBase: A Cross-Species Database for Comparative Genomics." *Nucleic Acids Research* 31(1) (2003): 133–37.

Harvey, Mark, and Andrew McMeekin. *Public or Private Economics of Knowledge? Turbulence in the Biological Sciences.* Cheltenham, UK: Edward Elgar, 2007.

Harwood, Jonathan. *Europe's Green Revolution and Others Since: The Rise and Fall of Peasant-Friendly Plant Breeding.* London: Routledge, 2012.

Helmreich, Stephen. *Silicon Second Nature: Culturing Artificial Life in a Digital World.* Berkeley, CA: University of California Press, 1998.

Hempel, Carl G. "Fundamentals of Concept Formation in Empirical Science." In *Foundations of the Unity of Science, Volume 2*, edited by Otto Neurath, Rudolf Carnap, and C. Morris, 651–746. Chicago: University of Chicago Press, 1970. Originally published as Carl G. Hempel, *Fundamentals of Concept Formation in Empirical Science* (Chicago: University of Chicago Press, 1952).

———. *Revolutions and Reconstructions in the Philosophy of Science.* Brighton, UK: Harvester, 1980.

Hesse, Mary. *The Structure of Scientific Inference.* London: Macmillan, 1974.

Hey, Tony, Stewart Tansley, and Kristine Tolle, eds. *The Fourth Paradigm: Data-Intensive Scientific Discovery.* Redmond, WA: Microsoft Research, 2009.

Hilgartner, Stephen. "Biomolecular Databases: New Communication Regimes for Biology?" *Science Communication* 17 (1995): 240–63.

———. "Constituting Large-Scale Biology: Building a Regime of Governance in the Early Years of the Human Genome Project." *BioSocieties* 8 (2013): 397–416.

———. "Mapping Systems and Moral Order: Constituting Property in Genome

Laboratories." In *States of Knowledge: The Co-Production of Science and Social Order*, edited by Sheila Jasanoff, 131–41. London: Routledge, 2004.

———. *Reordering Life: Knowledge and Control in the Genomic Revolution.* Cambridge, MA: MIT Press, 2016.

Hill, David P., Barry Smith, Monica S. McAndrews-Hill, and Judith A. Blake. "Gene Ontology Annotations: What They Mean and Where They Come From." *BMC Bioinformatics* 9(S5) (2008): S2.

Hinchliffe, Steve, and Stephanie Lavau. "Differentiated Circuits: The Ecologies of Knowing and Securing Life." *Environment and Planning D: Society and Space* 31(2) (2013): 259–74.

Hine, Christine. "Databases as Scientific Instruments and Their Role in the Ordering of Scientific Work." *Social Studies of Science* 36(2) (2006): 269–98.

———. *Systematics as Cyberscience: Computers, Change, and Continuity in Science.* Cambridge, MA: MIT Press, 2008.

Hinterberger, Amy, and Natalie Porter. "Genomic and Viral Sovereignty: Tethering the Materials of Global Biomedicine." *Public Culture* 27(2) (2015): 361–86.

Hoehndorf, Robert, Anika Oellrich, Michel Dumontier, Janet Kelso, Dietrich Rebholz-Schuhmann, and Heinrich Herre. "Relations as Patterns: Bridging the Gap between OBO and OWL." *BMC Bioinformatics* 11 (2010): 441.

Hong, Sungook. "The Hwang Scandal That 'Shook the World of Science.'" *East Asian Science, Technology and Society: An International Journal* 2 (2008): 1–7.

Howard, Don A. "Lost Wanderers in the Forest of Knowledge: Some Thoughts on the Discovery-Justification Distinction." In *Revisiting Discovery and Justification: Historical and Philosophical Perspectives on the Context Distinction*, edited by Jutta Schockore and Friedrich Steinle, 3–22. Dordrecht, Netherlands: Springer, 2006.

Howe, Doug, Maria Costanzo, Petra Fey, Takashi Gojobori, Linda Hannick, Winston Hide, David P. Hill et al. "Big Data: The Future of Biocuration." *Nature* 455 (2008): 47–50.

Hoyningen-Huene, Paul. *Systematicity: The Nature of Science.* Oxford: Oxford University Press, 2013.

Huala, Eva, Allan Dickerman, Margarita Garcia-Hernandez, Danforth Weems, Leonore Reiser, Frank LaFond, David Hanley et al. "The *Arabidopsis* Information Resource (TAIR): A Comprehensive Database and Web-Based Information Retrieval, Analysis and Visualization System for a Model Plant." *Nucleic Acids Research* 29(1) (2001): 102–5.

Huang, Shanjin, Robert C. Robinson, Lisa Y. Gao, Tracie Matsumoto, Arnaud Brunet, Laurent Blanchoin, and Christopher J. Staiger. "*Arabidopsis* VILLIN1 Generates Actin Filament Cables That Are Resistant to Depolymerization." *The Plant Cell* 17 (2005): 486–501.

Huber, Lara, and Lara K. Keuck. "Mutant Mice: Experimental Organisms as Materialised Models in Biomedicine." *Studies in History and Philosophy of Biological and Biomedical Sciences* 44(3) (2013): 385–91.

Hull, David. "Are Species Really Individuals?" *Systematic Zoology* 25(2) (1976): 173–91.

Hutchins, Edward. *Cognition in the Wild.* Cambridge, MA: MIT Press, 1995.

International Arabidopsis Informatics Consortium. "An International Bioinformatics Infrastructure to Underpin the *Arabidopsis* Community." *The Plant Cell* 22(8)
(2010): 2530–36.

———. "Taking the Next Step: Building an Arabidopsis Information Portal." *The
Plant Cell* 24(6) (2012): 2248–56.

Jasanoff, Sheila. "Making Order: Law and Science in Action." In *The Handbook
of Science and Technology Studies*, 3rd. ed., edited by Edward J. Hackett, Olga
Amsterdamska, Michael Lynch, and Judy Wajcman, 761–86. Cambridge, MA:
MIT Press, 2008.

———, ed. *States of Knowledge: The Co-Production of Science and the Social Order*. London: Routledge, 2004.

Jensen, Elaine Fiona. "Flowering Time Diversity in *Miscanthus*: A Tool for the Opti-
mization of Biomass." *Comparative Biochemistry and Physiology, Part A: Molecular and Integrative Physiology* 153(2) Supplement 1, S197 (2009). Supplement—
Abstracts of the Annual Main Meeting of the Society of Experimental Biology,
Glasgow, UK, June 28–July 1, 2009.

Jensen, Elaine Fiona, Kerrie Farrar, Sian Thomas Jones, Astley Hastings, Iain Simon Donnison, and John Cedric Clifton-Brown. "Characterization of Flowering
Time Diversity in *Miscanthus* Species." *GCB Bioenergy* 3 (2011): 387–400.

Johnson, Kristin. *Ordering Life: Karl Jordan and the Naturalist Tradition*. Baltimore:
Johns Hopkins University Press, 2012.

Jonkers, Koen. "Models and Orphans: Concentration of the Plant Molecular Life
Science Research Agenda." *Scientometrics* 83(1) (2010): 167–79.

JoVE. "About." Accessed October 17, 2015. http://www.jove.com/about.

Karaca, Koray. "A Study in the Philosophy of Experimental Exploration: The Case
of the ATLAS Experiment at CERN's Large Hadron Collider." *Synthese* (2016).

Kay, Lily E. *The Molecular Vision of Life: Caltech, the Rockefeller Foundation, and
the Rise of the New Biology*. Oxford: Oxford University Press, 1993.

———. *Who Wrote the Book of Life? A History of the Genetic Code*. Stanford, CA:
Stanford University Press, 2000.

Keet, Maria C. "Dependencies between Ontology Design Parameters." *International
Journal of Metadata, Semantics and Ontologies* 54 (2010): 265–84.

Kell, Douglas B., and Stephen G. Oliver. "Here Is the Evidence, Now What Is the Hypothesis? The Complementary Roles of Inductive and Hypothesis-Driven Science
in the Post-Genomic Era." *BioEssays* 26(1) (2004): 99–105.

Keller, Evelyn F. *The Century of the Gene*. Cambridge, MA: Harvard University
Press, 2000.

———. *A Feeling for the Organism: The Life and Work of Barbara McClintock*.
New York: W. H. Freeman, 1983.

Kellert, Stephen H., Helen E. Longino, and C. Kenneth Waters, eds. *Scientific Pluralism*. Minneapolis: University of Minnesota Press, 2006.

Kelty, Christopher M. "Outlaws, Hackers, Victorian Amateurs: Diagnosing Public
Participation in the Life Sciences Today." *Jcom* 9(01) (2010): C03.

———. "This Is Not an Article: Model Organism Newsletters and the Question of
'Open Science.'" *BioSocieties* 7(2) (2012): 140–68.

———. *Two Bits: The Cultural Significance of Free Software*. Durham, NC: Duke
University Press, 2008.

King, Ross D., Jem Rowland, Stephen G. Oliver, Michael Young, Wayne Aubrey, Emma Byrne, Maria Liakata et al. "The Automation of Science." *Science* 324 (2009): 85–89.

Kingsbury, Noel. *Hybrid: The History and Science of Plant Breeding*. Chicago: University of Chicago Press, 2009.

Kingsland, Sharon E. *Modeling Nature: Episodes in the History of Population Ecology*. Chicago: University of Chicago Press, 1995.

Kirk, Robert G. "A Brave New Animal for a Brave New World: The British Laboratory Animal Bureau and the Constitution of International Standards of Laboratory Animal Production and Use, Circa 1947–1968." *Isis* 101 (2010):62–94.

———. "'Standardization through Mechanization': Germ-Free Life and the Engineering of the Ideal Laboratory Animal." *Technology and Culture* 53 (2012): 61–93.

Kitchin, Rob. *The Data Revolution: Big Data, Open Data, Data Infrastructures and Their Consequences*. London: Sage, 2014.

Klein, Ursula, and Wolfgang Lefèvre. *Materials in Eighteenth-Century Science: A Historical Ontology*. Cambridge, MA: MIT Press, 2007.

Knorr Cetina, Karin D. *Epistemic Cultures: How the Sciences Make Knowledge*. Cambridge, MA: Harvard University Press, 1999.

Knowledge/Value. "Concept Note for the Workshop Series." Accessed October 17, 2015. http://www.knowledge-value.org.

Kohler, Robert E. *Lords of the Fly*: Drosophila *Genetics and the Experimental Life*. Chicago: University of Chicago Press, 1994.

Koornneef, Maarten, and David Meinke. "The Development of *Arabidopsis* as a Model Plant." *The Plant Journal* 61 (2010): 909–21.

Kourany, Janet A. *Philosophy of Science after Feminism*. Oxford: Oxford University Press, 2010.

Krakauer, David C., James P. Collins, Douglas Erwin, Jessica C. Flack, Walter Fontana, Manfred D. Laubichler, Sonja J. Prohaska et al. "The Challenges and Scope of Theoretical Biology." *Journal of Theoretical Biology* 276 (2011): 269–76.

Krohs, Ulrich. "Convenience Experimentation." *Studies in History and Philosophy of Biological and Biomedical Sciences* 43(1) (2012): 52–57.

Krohs, Ulrich, and Werner Callebaut. "Data without Models Merging with Models without Data." In *Systems Biology: Philosophical Foundations*, edited by Fred C. Boogerd, Frank J. Bruggeman, Jan-Hendrik S. Hofmeyr, and Hans V. Westerhoff, 181–213. Amsterdam: Elsevier, 2007.

Kuhn, Thomas S. *The Structure of Scientific Revolutions*. Chicago: University of Chicago Press, 1962.

Kusch, Martin. "Hacking's Historical Epistemology: A Critique of Styles of Reasoning." *Studies in History and Philosophy of Science* 41(2) (2010): 158–73.

———. "Scientific Pluralism and the Chemical Revolution." *Studies in the History and the Philosophy of Science: Part A* 49 (2015): 69–79.

Latour, Bruno. "Circulating Reference: Sampling the Soil in the Amazon Forest." In *Pandora's Hope: Essays on the Reality of Science Studies*, by Bruno Latour, 24–79. Cambridge, MA: Harvard University Press, 1999.

Latour, Bruno, and Steven Woolgar. *Laboratory Life: The Construction of Scientific Facts*. Princeton, NJ: Princeton University Press, 1979.

Lawrence, Rebecca. "Data: Why Openness and Sharing Are Important." *F1000-Research Blog*, March 14, 2013. Accessed October 17, 2015. http://blog.f1000research.com/2013/03/14/data-why-openness-and-sharing-are-important.

Ledford, Heidi. "Molecular Biology Gets Wikified." *Nature Online*, July 23, 2008. Accessed October 17, 2015. doi:10.1038/news.2008.971.

Lee, Charlotte P., Paul Dourish, and Gloria Mark. "The Human Infrastructure of Cyberinfrastructure." Paper presented at the Computer Supported Cooperative Work Conference (CSCW), Banff, Canada, November 4–8, 2006. Accessed October 17, 2015. http://www.dourish.com/publications/2006/cscw2006-cyberinfrastructure.pdf.

Lenoir, Timothy. "Shaping Biomedicine as an Information Science." In *Proceedings of the 1998 Conference on the History and Heritage of Science Information Systems*, edited by Mary Ellen Bowden et al., 27–45. Medford, NJ: Information Today, 1999.

Leonelli, Sabina. "Documenting the Emergence of Bio-Ontologies: Or, Why Researching Bioinformatics Requires HPSSB." *History and Philosophy of the Life Sciences* 32(1) (2010): 105–26.

———. "Epistemische Diversität im Zeitalter von Big Data: Wie Dateninfrastrukturen der biomedizinischen Forschung dienen." In *Diversität: Geschichte und Aktualität eines Konzepts*, edited by André Blum, Lina Zschocke, Hans-Jörg Rheinberger, and Vincent Barras, 85–106. Würzburg: Königshausen and Neumann, 2016.

———. "Global Data for Local Science: Assessing the Scale of Data Infrastructures in Biological and Biomedical Research." *BioSocieties* 8 (2013): 449–65.

———. "Growing Weed, Producing Knowledge. An Epistemic History of *Arabidopsis thaliana*." *History and Philosophy of the Life Sciences* 29(2) (2007): 55–87.

———. "The Impure Nature of Biological Knowledge." In *Scientific Understanding: Philosophical Perspectives*, edited by Henk de Regt, Sabina Leonelli, and Kai Eigner, 189–209. Pittsburgh, PA: University of Pittsburgh Press, 2009.

———. "Performing Abstraction. Two Ways of Modelling *Arabidopsis thaliana*." *Biology and Philosophy* 23(4) (2008): 509–28.

———. "Scientific Organisations as Social Movements: Reflections on How Social Theory Can Inform the Philosophy of Science." In *Festschrift Hans Radder*, edited by Henk W. de Regt and Chunglin Kwa, 39–52. Amsterdam, Netherlands: VU University Press, 2014.

———. *Weed for Thought. Using* Arabidopsis thaliana *to Understand Plant Biology*. PhD diss., Vrije Universiteit Amsterdam, 2007. Open Access. Accessed March 30, 2014. dare.ubvu.vu.nl/bitstream/1871/10703/1/7623.pdf.

———. "What Difference Does Quantity Make? On the Epistemology of Big Data in Biology." *Big Data and Society* 1 (2014):1–11.

———. "When Humans Are the Exception: Cross-Species Databases at the Interface of Clinical and Biological Research." *Social Studies of Science* 42(2) (2012): 214–36.

Leonelli, Sabina, and Rachel A. Ankeny. "Repertoires: How To Transform a Project into a Research Community." *BioScience* 65(7) (2015): 701–8.

———. "Re-Thinking Organisms: The Impact of Databases on Model Organism Biology." *Studies in History and Philosophy of Biological and Biomedical Sciences* 43(1) (2012): 29–36.

———. "What Makes a Model Organism?" *Endeavour* 37(4) (2013): 209–12.

Leonelli, Sabina, Rachel A. Ankeny, Nicole Nelson, and Edmund Ramsden. "Making Organisms Model Humans: Situated Models in Alcohol Research." *Science in Context* 27(3) (2014): 485–509.

Leonelli, Sabina, Alexander D. Diehl, Karen R. Christie, Midori A. Harris, and Jane Lomax. "How the Gene Ontology Evolves." *BMC Bioinformatics* 12 (2011): 325.

Leonelli, Sabina, Berris Charnley, Alex R. Webb, and Ruth Bastow. "Under One Leaf: An Historical Perspective on the UK Plant Science Federation." *New Phytologist* 195(1) (2012): 10–13.

Leonelli, Sabina, Brian Rappert, and Gail Davies. "Data Shadows: Knowledge, Openness and Access." *Science, Technology and Human Values*. Forthcoming.

Leonelli, Sabina, Nicholas Smirnoff, Jonathan Moore, Charis Cook, and Ruth Bastow. "Making Open Data Work for Plant Scientists." *Journal of Experimental Botany* 64(14) (2013): 4109–17.

Leonelli, Sabina, Daniel Spichtiger, and Barbara Prainsack. "Sticks AND Carrots: Incentives for a Meaningful Implementation of Open Science Guidelines." *Geo* 2 (2015): 12–16.

Levin, Nadine. "What's Being Translated in Translational Research? Making and Making Sense of Data Between the Laboratory and the Clinic." *Technoscienza* 5(1) (2014): 91–114.

Levin, Nadine, Dagmara Weckoswka, David Castle, John Dupré, and Sabina Leonelli. "How Do Scientists Understand Openness?" Under review.

Lewis, James, and Andrew Bartlett. "Inscribing a Discipline: Tensions in the Field of Bioinformatics." *New Genetics and Society* 32(3) (2013): 243–63.

Lewis, Suzanna E. "Gene Ontology: Looking Backwards and Forwards." *Genome Biology* 6(1) (2004): 103.

Liang, Chengzhi, Pankaj Jaiswal, Claire Hebbard, Shuly Avraham, Edward S. Buckler, Terry Casstevens, Bonnie Hurwitz et al. "Gramene: A Growing Plant Comparative Genomics Resource." *Nucleic Acids Research* 36 (2008): D947–D953.

Lipton, Peter. *Inference to the Best Explanation*. London: Routledge, 1991.

Livingstone, David N. *Putting Science in Its Place: Geographies of Scientific Knowledge*. Chicago: University of Chicago Press, 2003.

Logan, Cheryl A. "Before There Were Standards: The Role of Test Animals in the Production of Scientific Generality in Physiology." *Journal of the History of Biology* 35 (2002): 329–63.

———. "The Legacy of Adolf Meyer's Comparative Approach: Worcester Rats and the Strange Birth of the Animal Model." *Integrative Physiological and Behavioral Science* 40 (2005): 169–81.

Longino, Helen E. *The Fate of Knowledge*. Princeton, NJ: Princeton University Press, 2002.

———. *Science as Social Knowledge: Values and Objectivity in Scientific Inquiry*. Princeton, NJ: Princeton University Press, 1990.

Love, Alan C. "Typology Reconfigured: From the Metaphysics of Essentialism to the Epistemology of Representation." *Acta Biotheoretica* 57 (2009): 51–75.

Love, Alan C., and Michael Travisano. "Microbes Modeling Ontogeny." *Biology and Philosophy* 28(2) (2013): 161–88.

Lynch, Michael. "Protocols, Practices, and the Reproduction of Technique in Molecular Biology." *British Journal of Sociology* 53(2) (2002): 203–20.

Mackenzie, Adrian. "Bringing Sequences to Life: How Bioinformatics Corporealises Sequence Data." *New Genetics and Society* 22(3) (2003): 315–32.

MacLean, Dan, Kentaro Yoshida, Anne Edwards, Lisa Crossman, Bernardo Clavijo, Matt Clark, David Swarbreck et al. "Crowdsourcing Genomic Analyses of Ash and Ash Dieback—Power to the People." *GigaScience* 2 (2013): 2. Accessed October 17, 2015. http://www.gigasciencejournal.com/content/2/1/2.

MacLeod, Miles, and Nancy Nersessian. "Building Simulations from the Ground Up: Modeling and Theory in Systems Biology." *Philosophy of Science* 80 (2013): 533–56.

Maddox, Brenda. *Rosalind Franklin: The Dark Lady of DNA*. New York: HarperCollins, 2002.

Maher, Brendan. "Evolution: Biology's Next Top Model?" *Nature* 458 (2009): 695–98.

Malone, James, and Robert Stevens. "Measuring the Level of Activity in Community Built Bio-Ontologies." *Journal of Biomedical Informatics* 46 (2013): 5–14.

Martin, Paul. "Genetic Governance: The Risks, Oversight and Regulation of Genetic Databases in the UK." *New Genetics and Society* 20(2) (2001): 157–83.

Massimi, Michela. "From Data to Phenomena: A Kantian Stance." *Synthese* 182 (2009): 101–16.

Maxson, Kathryn M., Robert Cook-Deegan, and Rachel A. Ankeny. *The Bermuda Triangle: Principles, Practices, and Pragmatics in Genomic Data Sharing*. In preparation.

Mayer-Schönberger, Viktor, and Kenneth Cukier. *Big Data: A Revolution that Will Transform How We Live, Work and Think*. London: John Murray, 2013.

Mazzotti, Massimo. "Lessons from the L'Aquila Earthquake." *Times Higher Education*, October 3, 2013. Accessed January 22, 2014. http://www.timeshigher education.co.uk/features/lessons-from-the-laquila-earthquake/2007742.fullarticle.

McAllister, James W. "Model Selection and the Multiplicity of Patterns in Empirical Data." *Philosophy of Science* 74(5) (2007): 884–94.

———. "The Ontology of Patterns in Empirical Data." *Philosophy of Science* 77(5) (2010): 804–14.

———. "Phenomena and Patterns in Data Sets." *Erkenntnis* 47 (1997): 217–28.

———. "What Do Patterns in Empirical Data Tell Us About the Structure of the World?" *Synthese* 182 (2011): 73–87.

McCain, Katherine W. "Mandating Sharing: Journal Policies in the Natural Sciences." *Science Communication* 16(4) (1995): 403–31.

McCarthy, Mark I., Gonçalo R. Abecasis, Lon R. Cardon, David B. Goldstein, Julian Little, John P. A. Ioannidis, and Joel N. Hirschhorn. "Genome-Wide Association Studies for Complex Traits: Consensus, Uncertainty and Challenges." *Nature Reviews Genetics* 9(5) (2009): 356–69.

McMullen, Patrick D., Richard I. Morimoto, and Luís A. Nunes Amaral. "Physically Grounded Approach for Estimating Gene Expression from Microarray Data." *Proceedings of the National Academy of Sciences of the United States of America* 107(31) (2010): 13690–695.

McMullin, Ernan. "Values in Science." *Philosophy of Science* 4 (1982): 3–28.

McNally, Ruth, Adrian Mackenzie, Allison Hui, and Jennifer Tomomitsu. "Understanding the 'Intensive' in 'Data Intensive Research': Data Flows in Next Generation Sequencing and Environmental Networked Sensors." *International Journal of Digital Curation* 7(1) (2012): 81–95.

Mill, John Stuart. *A System of Logic, Ratiocinative and Inductive: Being a Connected View of the Principles of Evidence, and the Methods of Scientific Investigation.* London: John W. Parker, 1843.

Minelli, Alessandro. *The Development of Animal Form.* Cambridge: Cambridge University Press, 2003.

Mitchell, Sandra D. *Biological Complexity and Integrative Pluralism.* Cambridge: Cambridge University Press, 2003.

———. *Unsimple Truths: Science, Complexity, and Policy.* Chicago: University of Chicago Press, 2009.

Mobley, Aaron, Suzanne K. Linder, Russell Braeuer, Lee M. Ellis, and Leonard Zwelling. "A Survey on Data Reproducibility in Cancer Research Provides Insights into Our Limited Ability to Translate Findings from the Laboratory to the Clinic." *PLOS ONE* 8(5) (2013): e63221.

Moody, Glyn. *Digital Code of Life: How Bioinformatics is Revolutionizing Science, Medicine and Business.* Hoboken, NJ: Wiley, 2004.

Morange, Michel. *A History of Molecular Biology.* Cambridge, MA: Harvard University Press, 1998.

Morgan, Mary S. "Experiments without Material Intervention: Model Experiments, Virtual Experiments, and Virtually Experiments." In *The Philosophy of Scientific Experimentation*, edited by Hans Radder, 216–35. Pittsburgh, PA: University of Pittsburgh Press, 2003.

———. "Resituating Knowledge: Generic Strategies and Case Studies." *Philosophy of Science* 81 (2014): 1012–24.

———. "Travelling Facts." In *How Well Do Facts Travel?: The Dissemination of Reliable Knowledge*, edited by Peter Howlett and Mary S. Morgan, 3–42. Cambridge: Cambridge University Press, 2010.

———. *The World in the Model: How Economists Work and Think.* Cambridge: Cambridge University Press, 2012.

Morgan, Mary S., and Margaret Morrison. *Models as Mediators: Perspectives on Natural and Social Science.* Cambridge: Cambridge University Press, 1999.

Morrison, Margaret. *Unifying Scientific Theories: Physical Concepts and Mathematical Structures.* Cambridge: Cambridge University Press, 2007.

Mueller, Lukas A., Peifen Zhang, and Seung Y. Rhee. "AraCyc: A Biochemical Pathway Database for Arabidopsis." *Plant Physiology* 132 (2003): 453–60.

Müller-Wille, Staffan. "Collection and Collation: Theory and Practice of Linnaean Botany." *Studies in History and Philosophy of Biological and Biomedical Sciences* 38 (2007): 541–62.

Müller-Wille, Staffan, and Isabelle Charmantier. "Lists as Research Technologies." *Isis* 103(4) (2012): 743–52.

———. "Natural History and Information Overload: The Case of Linnaeus." *Studies in History and Philosophy of Biological and Biomedical Sciences* 43 (2012): 4–15.

Müller-Wille, Staffan, and James Delbourgo. "LISTMANIA: Introduction." *Isis* 103(4) (2012): 710–15.

Müller-Wille, Staffan, and Hans-Jörg Rheinberger. *A Cultural History of Heredity.* Chicago: University of Chicago Press, 2012.

National Cancer Institute. "An Assessment of the Impact of the NCI Cancer Biomedical Informatics Grid (caBIG®)." Report published March 3, 2011. Accessed October 17, 2015. http://deainfo.nci.nih.gov/advisory/bsa/archive/bsa0311/caBIG finalReport.pdf.

National Human Genome Research Institute. "Genome Informatics and Computational Biology Program." Accessed January 14, 2014. http://www.genome.gov /10001735.

———. "Model Organism Databases Supported by the National Human Genome Research Institute." Accessed October 17, 2015. http://www.genome.gov /10001837.

Nature. "Challenges in Irreproducible Research (Special)." 2013. Accessed October 17, 2015. http://go.nature.com/huhbyr.

Nature. "The Future of Publishing (Special)." 2013. Accessed October 17, 2015. http://www.nature.com/news/specials/scipublishing/index.html.

Nelson, Nicole C. "Modeling Mouse, Human and Discipline: Epistemic Scaffolds in Animal Behavior Genetics." *Social Studies of Science* 43(1) (2013): 3–29.

Nisen, Perry, and Frank Rockhold. "Access to Patient-Level Data from Glaxo SmithKline Clinical Trials." *New England Journal of Medicine* 369(5) (2013): 475–78.

Nordmann, Alfred, Hans Radder, and Gregor Schiemann, eds. *Science Transformed? Debating Claims of an Epochal Break.* Pittsburgh, PA: University of Pittsburgh Press, 2011.

November, Joseph A. *Biomedical Computing: Digitizing Life in the United States.* Baltimore: Johns Hopkins University Press, 2012.

O'Malley, Maureen A. "Exploration, Iterativity and Kludging in Synthetic Biology." *Comptes Rendus Chimie* 14(4) (2011): 406–12.

———. "Exploratory Experimentation and Scientific Practice: Metagenomics and the Proteorhodopsin Case." *History and Philosophy of the Life Sciences* 29(3) (2008): 337–58.

———. *Philosophical Issues in Microbiology.* Cambridge: Cambridge University Press, 2014.

O'Malley, Maureen A., Kevin C. Elliott, Chris Haufe, and Richard M. Burian. "Philosophies of Funding." *Cell* 138 (2009): 611–15.

O'Malley, Maureen A., and Orkun S. Soyer. "The Roles of Integration in Molecular Systems Biology." *Studies in the History and the Philosophy of the Biological and Biomedical Sciences* 43(1) (2012): 58–68.

O'Malley, Maureen A., and Karola Stotz. "Intervention, Integration and Translation in Obesity Research: Genetic, Developmental and Metaorganismal Approaches." *Philosophy, Ethics, and Humanities in Medicine* 6 (2011): 2.

Open Science Collaborative. "Estimating the Reproducibility of Psychological Science." *Science* 349(6251) (2015).

Organisation for Economic Co-Operation and Development. "Guidelines for Human Biobanks and Genetic Research Databases (HBGRDs)." 2009. Accessed October 17, 2015. http://www.oecd.org/sti/biotechnology/hbgrd.

———. "OECD Principles and Guidelines for Access to Research Data from Public Funding." 2007. Accessed October 17, 2015. http://www.oecd.org/science/sci-tech/38500813.pdf.

Ort, Donald R., and Aleel K. Grennan. "*Plant Physiology* and TAIR Partnership." *Plant Physiology* 146 (2008): 1022–23.

Oyama, Susan. *The Ontogeny of Information: Developmental Systems and Evolution*. Cambridge: Cambridge University Press, 2000.

Paracer, Surindar, and Vernon Ahmadjian. *Symbiosis: An Introduction to Biological Associations*. Oxford: Oxford University Press, 2000.

Parker, John N., Niki Vermeulen, and Bart Penders. *Collaboration in the New Life Sciences*. Farnham, UK: Ashgate, 2010.

Parker, Wendy. "Does Matter Really Matter? Computer Simulations, Experiments, and Materiality." *Synthese* 169(3) (2009): 483–96.

Penders, Bart, Klasien Horstman, and Rein Vos. "Walking the Line Between Lab and Computation: The 'Moist' Zone." *BioScience* 58(8) (2008): 747–55.

Perspectives on Science. "Philosophy of Science after Feminism (Special Section)." *Perspectives on Science* 20(3) (2012).

Pickstone, John V. *Ways of Knowing: A New History of Science, Technology and Medicine*. Manchester: Manchester University Press, 2000.

Piwowar, Heather. "Who Shares? Who Doesn't? Factors Associated with Openly Archiving Research Data." *PLOS ONE* 6(7) (2007): e18657.

Polanyi, Michael. *Personal Knowledge: Towards a Post-Critical Philosophy*. Chicago: University of Chicago Press, 1958.

Pollock, Kevin, Adel Fakhir, Zoraida Portillo, Madhukara Putty, and Paula Leighton. *The Hidden Digital Divide*. Accessed October 17, 2015. http://www.scidev.net/global/icts/data-visualisation/digital-divide-data-interactive.html.

Popper, Karl R. "Models, Instruments, and Truth." In *The Myth of the Framework: In Defence of Science and Rationality*, edited by M. A. Notturno, 154–84. London: Routledge, 1994.

Porter, Theodore M. *Trust in Numbers: The Pursuit of Objectivity in Science and Public Life*. Princeton, NJ: Princeton University Press, 1995.

Potter, Clive, Tom Harwood, Jon Knight, and Isobel Tomlinson. "Learning from History, Predicting the Future: The UK Dutch Elm Disease Outbreak in Relation to Contemporary Tree Disease Threats." *Philosophical Transactions of the Royal Society B* 366(1573) (2011): 1966–74.

Powledge, Tabitha M. "Changing the Rules?" *EMBO Reports* 2(3) (2001): 171–72.

Prainsack, Barbara. *Personalization from Below: Participatory Medicine in the 21st Century*. New York: New York University Press, forthcoming.

Prainsack, Barbara, and Sabina Leonelli. "To What Are We Opening Science?" *LSE Impact Blog*. Published online May 2015. http://blogs.lse.ac.uk/impactofsocial sciences/2015/04/21/to-what-are-we-opening-science.

Princeton University. "John Bonner's Slime Mold Movies." Accessed April 9, 2014. http://www.youtube.com/watch?v=bkVhLJLG7ug.

Quirin, Edmund A., Harpartap Mann, Rachel S. Meyer, Alessandra Traini, Maria

Luisa Chiusano, Amy Litt, and James M. Bradeen. "Evolutionary Meta-Analysis of Solanaceous Resistance Gene and *Solanum* Resistance Gene Analog Sequences and a Practical Framework for Cross-Species Comparisons." *Molecular Plant-Microbe Interactions* 25(5) (2012): 603–12.

Radder, Hans. *The Material Realization of Science: From Habermas to Experimentation and Referential Realism.* Dordrecht, Netherlands: Springer, 2012.

———. *The Philosophy of Scientific Experimentation.* Pittsburgh, PA: University of Pittsburgh Press, 2003.

———. "The Philosophy of Scientific Experimentation: A Review." *Automated Experimentation* 1 (2009): 2.

———. *The World Observed/The World Conceived.* Pittsburgh, PA: University of Pittsburgh Press, 2006.

Rader, Karen A. *Making Mice: Standardizing Animals for American Biomedical Research, 1900–1955.* Princeton, NJ: Princeton University Press, 2004.

Raj, Kapil. *Relocating Modern Science: Circulation and the Construction of Knowledge in South Asia and Europe.* Delhi: Palgrave Macmillan, 2006.

Ramsden, Edmund. "Model Organisms and Model Environments: A Rodent Laboratory in Science, Medicine and Society." *Medical History* 55 (2011): 365–68.

Rappert, Brian, and Brian Balmer, eds. *Absence in Science, Security and Policy: From Research Agendas to Global Strategies.* New York: Palgrave Macmillan, 2015.

Reichenbach, Hans. *Experience and Prediction: An Analysis of the Foundations and the Structure of Knowledge.* Chicago: University of Chicago Press, 1938.

Renear, Allen H., and Carole L. Palmer. "Strategic Reading, Ontologies, and the Future of Scientific Publishing." *Science* 325(5942) (2009): 828–32.

Research Councils UK. "RCUK Policy on Open Access." 2012. Accessed October 17, 2015. http://www.rcuk.ac.uk/research/openaccess/policy.

Resnik, David B. *Owning the Genome: A Moral Analysis of DNA Patenting.* Albany, NY: State University of New York Press, 2004.

Reydon, Thomas A. "Natural Kind Theory as a Tool for Philosophers of Science." In *EPSA Epistemology and Methodology of Science: Launch of the European Philosophy of Science Association*, edited by Mauricio Suárez, Mauro Dorato, and Miklós Rédei, 245–54. Dordrecht, Netherlands: Springer, 2010.

Rhee, Seung Yon. "Carpe Diem. Retooling the 'Publish or Perish' Model into the 'Share and Survive' Model." *Plant Physiology* 134(2) (2004): 543–47.

Rhee, Seung Yon, William Beavis, Tanya Z. Berardini, Guanghong Chen, David Dixon, Aisling Doyle, Margarita Garcia-Hernandez, Eva Huala, Gabriel Lander et al. "The *Arabidopsis* Information Resource (TAIR): A Model Organism Database Providing a Centralized, Curated Gateway to *Arabidopsis* Biology, Research Materials and Community." *Nucleic Acids Research* 31(1) (2003): 224–28.

Rhee, Seung Yon, and Bill Crosby. "Biological Databases for Plant Research." *Plant Physiology* 138 (2005): 1–3.

Rhee, Seung Yon, Julie Dickerson, and Dong Xu. "Bioinformatics and Its Applications in Plant Biology." *Annual Review of Plant Biology* 57 (2006): 335–60.

Rhee, Seung Yon, Valerie Wood, Kara Dolinski, and Sorin Draghici. "Use and Misuse of the Gene Ontology (GO) Annotations." *Nature Reviews Genetics* 9(7) (2008): 509–15.

Rheinberger, Hans-Jörg. *An Epistemology of the Concrete*. Durham, NC: Duke University Press, 2010.

———. "Infra-Experimentality: From Traces to Data, from Data to Patterning Facts." *History of Science* 49(3) (2011): 337–48.

———. *Towards a History of Epistemic Things: Synthesizing Proteins in the Test Tube*. Redwood City, CA: Stanford University Press, 1997.

Richardson, Michael K., James Hanken, Mayoni L. Gooneratne, Claude Pieau, Albert Raynaud, Lynne Selwood, and Glenda M. Wright. "There Is No Highly Conserved Embryonic Stage in the Vertebrates: Implications for Current Theories of Evolution and Development." *Anatomy and Embryology* 196(2) (1997): 91–106.

Rizzo, Dan M., et al. "*Phytophthora ramorum* and Sudden Oak Death in California 1: Host Relationships." In *Proceedings of the Fifth Symposium on Oak Woodlands: Oak Woodlands in California's Changing Landscape* (October 22–25, 2001, San Diego, CA). USDA Forest Service General Technical Report PSW-GTR-184 (February 2002). Albany, CA: Pacific Southwest Research Station, Forest Service, U.S. Department of Agriculture.

Rogers, Susan, and Alberto Cambrosio. "Making a New Technology Work: The Standardization and Regulation of Microarrays." *Yale Journal of Biology and Medicine* 80 (2007): 165–78.

Rooney, Phyllis. "On Values in Science: Is the Epistemic/Non-Epistemic Distinction Useful?" *Philosophy of Science* 1 (1992): 13–22.

Rosenthal, Nadia, and Michael Ashburner. "Taking Stock of Our Models: The Function and Future of Stock Centres." *Nature Reviews Genetics* 3 (2002): 711–17.

Rouse, Joseph. *Knowledge and Power: Toward a Political Philosophy of Science*. Ithaca, NY: Cornell University Press, 1987.

Royal Society. "Science as an Open Enterprise." Accessed October 17, 2015. http://royalsociety.org/policy/projects/science-public-enterprise/report.

Russell, Bertrand. *The Basic Writings of Bertrand Russell*. New York: Routledge, 2009.

Ryle, Gilbert. *The Concept of Mind*. Chicago: University of Chicago Press, 1949.

Sanger, Fred. "Sequences, Sequences and Sequences." *Annual Review of Biochemistry* 57 (1988): 1–29.

Schaeper, Nina D., Nikola-Michael Prpic, and Ernst A. Wimmer. "A Clustered Set of Three Sp-Family Genes Is Ancestral in the Metazoa." *BMC Evolutionary Biology* 10 (2010): 88.

Schaffer, Simon, Lissa Roberts, Kapil Raj, and James Delbourgo, eds. *The Brokered World: Go-Betweens and Global Intelligence, 1770–1820*. Sagamore Beach, MA: Watson Publishing International, 2009.

Schaffner, Kenneth F. *Discovery and Explanation in Biology and Medicine*. Chicago: University of Chicago Press, 1993.

Schickore, Jutta. "Studying Justificatory Practice: An Attempt to Integrate the History and Philosophy of Science." *International Studies in the Philosophy of Science* 23(1) (2009): 85–107.

Schickore, Jutta, and Friedrich Steinle, eds. *Revisiting Discovery and Justification: Historical and Philosophical Perspectives on the Context Distinction*. Dordrecht, Netherlands: Springer, 2006.

Schindler, Samuel. "Bogen and Woodward's Data/Phenomena Distinction: Forms of Theory-Ladenness, and the Reliability of Data." *Synthese* 182(1) (2011): 39–55.

———. "Theory-Laden Experimentation." *Studies in History and Philosophy of Science* 44(1) (2013): 89–101.

Scriven, Michael. "Explanation in the Biological Sciences." *Journal of the History of Biology* 2(1) (1969): 187–98.

Scudellari, Megan. "Data Deluge." *The Scientist*, October 1, 2011. Accessed October 17, 2015. http://www.the-scientist.com/?articles.view/articleNo/31212/title/Data-Deluge.

Searls, David B. "The Roots of Bioinformatics." *PLOS Computational Biology* 6(6) (2010): e1000809.

Secord, James. "Knowledge in Transit." *Isis* 95(4) (2004): 654–72.

Sepkoski, David. *Rereading the Fossil Record: The Growth of Paleobiology as an Evolutionary Discipline.* Chicago: University of Chicago Press, 2012.

Sepkoski, David, Elena Aronova, and Christine van Oertzen. "Historicizing Big Data." Special issue, *Osiris* (2017).

Shapin, Steven, and Simon Schaffer. *Leviathan and the Air-Pump: Hobbes, Boyle and the Experimental Life.* Princeton, NJ: Princeton University Press, 1985.

Shavit, Ayelet, and James R. Griesemer. "Transforming Objects into Data: How Minute Technicalities of Recording 'Species Location' Entrench a Basic Challenge for Biodiversity." In *Science in the Context of Application*, edited by Martin Carrier and Alfred Nordmann, 169–93. Boston: Boston Studies in the Philosophy of Science, 2011.

Shields, Robert. "MIAME, We Have a Problem." *Trends in Genetics* 22(2) (2006): 65–66.

Sipper, Moshe. *Machine Nature: The Coming Age of Bio-Inspired Computing.* Cambridge, MA: MIT Press, 2002.

Sivasundaram, Sujit. "Sciences and the Global: On Methods, Questions and Theory." *Isis* 101(1) (2010): 146–58.

Smith, Barry, Michael Ashburner, Cornelius Rosse, Jonathan Bard, William Bug, Werner Ceusters, Louis J. Goldberg et al. "The OBO Foundry: Coordinated Evolution of Ontologies to Support Biomedical Data Integration." *Nature Biotechnology* 25(11) (2007): 1251–55.

Smith, Barry, and Werner Ceusters. "Ontological Realism: A Methodology for Coordinated Evolution of Scientific Ontologies." *Applied Ontology* 5 (2010): 139–88.

Smith, Barry, Werner Ceusters, Bert Klagges, Jacob Köhler, Anand Kumar, Jane Lomax, Chris Mungall et al. "Relations in Biomedical Ontologies." *Genome Biology* 6 (2005): R46.

Smocovitis, Vassiliki Betty. "Keeping Up with Dobzhansky: G. Ledyard Stebbins, Jr., Plant Evolution, and the Evolutionary Synthesis." *History and Philosophy of the Life Sciences* 28(1) (2006): 9–47.

Somerville, Chris, and Maarten Koornneef. "A Fortunate Choice: The History of *Arabidopsis* as a Model Plant." *Nature Reviews Genetics* 3 (2002): 883–89.

Sommer, Ralf J. "The Future of Evo-Devo: Model Systems and Evolutionary Theory." *Nature Reviews Genetics* 10 (2009): 416–22.

Spradling, Allan, Barry Ganetsky, Phil Hieter, Mark Johnston, Maynard Olson, Terry Orr-Weaver, Janet Rossant et al. "New Roles for Model Genetic Organisms in Understanding and Treating Human Disease: Report From The 2006 Genetics Society of America Meeting." *Genetics* 172(4) (2006): 2025–32.

Sprague, Judy, Leyla Bayraktaroglu, Dave Clements, Tom Conlin, David Fashena, Ken Frazer, Melissa Haendel et al. "The Zebrafish Information Network: The Zebrafish Model Organism Database." *Nucleic Acids Research* 34(1) (2006): D581–D585.

Star, Susan L. "The Ethnography of Infrastructure." *American Behavioral Scientist* 43(3) (1999): 377–91.

Star, Susan L., and James R. Griesemer. "Institutional Ecology, 'Translations' and Boundary Objects: Amateurs and Professionals in Berkeley's Museum of Vertebrate Zoology, 1907–39." *Social Studies of Science* 19(3) (1989): 387–420.

Star, Susan L., and Katherine Ruhleder. "Steps Toward an Ecology of Infrastructure: Design and Access for Large Information Spaces." *Information Systems Research* 7(1) (1996): 111–34.

Stein, Lincoln D. "Towards a Cyberinfrastructure for the Biological Sciences: Progress, Visions and Challenges." *Nature Reviews Genetics* 9(9) (2008): 678–88.

Steinle, Friedrich. "Entering New Fields: Exploratory Uses of Experimentation." *Philosophy of Science* 64 (1997): S65–S74.

Stemerding, Dirk, and Stephen Hilgartner. "Means of Coordination in Making Biological Science: On the Mapping of Plants, Animals, and Genes." In *Getting New Technologies Together: Studies in Making Sociotechnical Order*, edited by Cornelis Disco and Barend van der Meulen, 39–70. Berlin: Walter de Gruyter, 1998.

Stevens, Hallam. *Life Out of Sequence: Bioinformatics and the Introduction of Computers into Biology*. Chicago: University of Chicago Press, 2013.

Stevens, Hallam, and Sarah Richardson, eds. *Postgenomics*. Durham, NC: Duke University Press, 2015.

Stitt, Mark, John Lunn, and Björn Usadel. "*Arabidopsis* and Primary Photosynthetic Metabolism—More than the Icing on the Cake." *The Plant Journal* 61(6) (2010): 1067–91.

Stotz, Karola, Paul E. Griffiths, and Rob Knight. "How Biologists Conceptualize Genes: An Empirical Study." *Studies in History and Philosophy of Biological and Biomedical Sciences* 35(4) (2004): 647–73.

Strasser, Bruno J. "The Experimenter's Museum: GenBank, Natural History, and the Moral Economies of Biomedicine." *Isis* 102 (2011): 60–96.

———. "GenBank—Natural History in the 21st Century?" *Science* 322(5901) (2008): 537–38.

———. "Laboratories, Museums, and the Comparative Perspective: Alan A. Boyden's Quest for Objectivity in Serological Taxonomy, 1924–1962." *Historical Studies in the Natural Sciences* 40(2) (2010): 149–82.

Suárez, Mauricio, and Nancy Cartwright. "Theories: Tools versus Models." *Studies in History and Philosophy of Modern Physics* 39(1) (2008): 62–81.

Suárez-Díaz, Edna. "The Rhetoric of Informational Molecules: Authority and Promises in the Early Study of Molecular Evolution." *Science in Context* 20(4) (2007): 649–77.

Suárez-Díaz, Edna, and Victor H. Anaya-Muñoz. "History, Objectivity, and the Construction of Molecular Phylogenies." *Studies in the History and Philosophy of Biological and Biomedical Sciences* 39(4) (2008): 451–68.

Subrahmanyam, Sanjay. *Explorations in Connected History, Vols. 1 and 2*. New Delhi: Oxford University Press, 2005.

Sunder Rajan, Kaushik. *Biocapital: The Constitution of Postgenomic Life*. Durham, NC: Duke University Press, 2006.

———. *Pharmocracy: Value, Politics and Knowledge in Global Biomedicine*. Durham, NC: Duke University Press, 2016.

Sunder Rajan, Kaushik, and Sabina Leonelli. "Introduction: Biomedical Trans-Actions, Postgenomics and Knowledge/Value." *Public Culture* 25(3) (2013): 463–75.

Suppes, Patrick. "From Theory to Experiment and Back Again." In *Observation and Experiment in the Natural and Social Sciences*, edited by Maria C. Galavotti, 1–41. Dordrecht, Netherlands: Kluwer, 2003.

———. "The Future Role of Computation in Science and Society." In *New Directions in the Philosophy of Science*, edited by Maria C. Galavotti et al., 35–44. Cham, Switzerland: Springer, 2014.

———. "Models of Data." In *Logic, Methodology and Philosophy of Science: Proceedings of the 1960 International Congress*, edited by Ernest Nagel, Patrick Suppes, and Alfred Tarski, 252–61. Stanford, CA: Stanford University Press, 1962. Reprinted in Patrick Suppes, *Studies in the Methodology and Foundations of Science. Selected Papers from 1951 to 1969*, 24–35. Dordrecht, Netherlands: Reidel, 1969.

———. "Perception, Models and Data: Some Comments." *Behavior Research Methods, Instruments and Computers* 29(1) (1997): 109–12.

———. "Statistical Concepts in Philosophy of Science." *Synthese* 154 (2007): 498–96.

Synaptic Leap. "Homepage." Accessed October 17, 2015. http://www.thesynaptic leap.org.

Tauber, Alfred I., and Sahotra Sarkar. "The Human Genome Project: Has Blind Reductionism Gone Too Far?" *Perspectives in Biology and Medicine* 35 (1992): 220–35.

Taylor, Chris F., Dawn Field, Susanna-Assunta Sansone, Jan Aerts, Rolf Apweiler, Michael Ashburner, Catherine A. Ball et al. "Promoting Coherent Minimum Reporting Guidelines for Biological and Biomedical Investigations: The MIBBI Project." *Nature Biotechnology* 26(8) (2008): 889–96.

Teller, Paul. "Saving the Phenomena Today." *Philosophy of Science* 77(5) (2010): 815–26.

Timpson, Christopher G. *Quantum Information Theory and the Foundations of Quantum Mechanics*. Oxford: Oxford University Press, 2013.

Tirmizi, Syed H., Stuart Aitken, Dilvan A. Moreira, Chris Mungall, Juan Sequeda, Nigam H. Shah, and Daniel P. Miranker. "Mapping between the OBO and OWL Ontology Languages." *Journal of Biomedical Semantics* 2(S1) (2011): S3.

Todes, Daniel P. *Pavlov's Physiology Factory: Experiment, Interpretation, Laboratory Enterprise*. Baltimore: Johns Hopkins University Press, 2001.

Tutton, Richard, and Barbara Prainsack. "Enterprising or Altruistic Selves? Making Up Research Subjects in Genetics Research." *Sociology of Health and Illness* 33(7) (2011): 1081–95.

Ure, Jenny, Rob Procter, Yu-wei Lin, Mark Hartswood, and Kate Ho. "Aligning Technical and Human Infrastructures in the Semantic Web: A Socio-Technical Perspective." Paper presented at the Third International Conference on e-Social

Science, University of Michigan, Ann Arbor, MI, October 7–9, 2007. Accessed October 17, 2015. http://citeseerx.ist.psu.edu/viewdoc/download?doi=10.1.1.97 .4584&rep=rep1&type=pdf.

Vakarelov, Orlin K. "The Information Medium." *Philosophy and Technology* 25(1) (2012): 47–65.

van Ommen, Gert-Jan B. "Popper Revisited: GWAS Here, Last Year." *European Journal of Human Genetics* 16 (2008): 1–2.

Vickers, John. "The Problem of Induction." In *The Stanford Encyclopedia of Philosophy*, edited by Edward N. Zalta et al. Accessed October 17, 2015. http://plato .stanford.edu/entries/induction-problem.

Waters, C. Kenneth. "How Practical Know-How Contextualizes Theoretical Knowledge: Exporting Causal Knowledge from Laboratory to Nature." *Philosophy of Science* 75 (2008): 707–19.

———. "The Nature and Context of Exploratory Experimentation." *History and Philosophy of the Life Sciences* 29 (2007): 1–9.

———. "What Concept Analysis in Philosophy Should Be (and Why Competing Philosophical Analyses of Gene Concepts Cannot Be Tested By Polling Scientists)." *History and Philosophy of the Life Sciences* 26 (2004): 29–58.

Weber, Marcel. *Philosophy of Experimental Biology*. Cambridge: Cambridge University Press, 2005.

Weisberg, Michael. *Simulation and Similarity: Using Models to Understand the World*. New York: Oxford University Press, 2013.

Wellcome Trust. "Sharing Data from Large-Scale Biological Research Projects: A System of Tripartite Responsibility." Report of a meeting organized by the Wellcome Trust, Fort Lauderdale, Florida, 14–15 January, 2003. Accessed October 17, 2015. http://www.genome.gov/Pages/Research/WellcomeReport0303.pdf.

Whewell, William. "Novum Organon Renovatum, Book II," In *William Whewell's Theory of Scientific Method*, edited by Robert E. Butts, 103–249. Indianapolis, IN: Hackett, 1989. Originally published in William Whewell, *Novum Organon Renovatum, Book II* (London: John W. Parker, 1858).

Wimsatt, William C. *Re-Engineering Philosophy for Limited Beings: Piecewise Approximations to Reality*. Cambridge, MA: Harvard University Press, 2007.

Wimsatt, William C., and James R. Griesemer. "Reproducing Entrenchments to Scaffold Culture: The Central Role of Development in Cultural Evolution." In *Integrating Evolution and Development: From Theory to Practice*, edited by Roger Sansom and Robert Brandon, 227–323. Cambridge, MA: MIT Press, 2007.

Winther, Rasmus G. "The Structure of Scientific Theories." In *The Stanford Encyclopedia of Philosophy*, edited by Edward N. Zalta et al. Accessed October 17, 2015. http://plato.stanford.edu/entries/structure-scientific-theories.

Woodward, James. "Data, Phenomena, and Reliability." *Philosophy of Science* 67 (2000): S163–S179.

———. "Phenomena, Signal, and Noise." *Philosophy of Science* 77 (2010): 792–803.

World Economic Forum. "Global Risks 2011: Sixth Edition." Accessed October 17, 2015. http://reports.weforum.org/global-risks-2011.

Wouters, Paul, and Colin Reddy. "Big Science Data Policies." In *Promise and Practice in Data Sharing*, edited by Paul Wouters and Peter Schröder. Amsterdam, Netherlands: Nederlands Instituut voor Wetenschappelijke Informatiediensten &

Koninklijke Nederlandse Akademie van Wetenschappen (NIWI-KNAW), 2003. Accessed October 17, 2015. http://virtualknowledgestudio.nl/documents/public -domain.pdf.

Wouters, Paul, Anne Beaulieu, Andrea Scharnhorst, and Sally Wyatt, eds. *Virtual Knowledge: Experimenting in the Humanities and the Social Sciences.* Cambridge, MA: MIT Press, 2013.

Wylie, Alison. *Thinking from Things: Essays in the Philosophy of Archaeology.* Berkeley, CA: University of California Press, 2002.

Wylie, Alison, Harold Kinkaid, and John Dupré. *Value-Free Science? Ideals and Illusions.* Oxford: Oxford University Press, 2007.

Index

Page numbers in italics refer to illustrations.